A Taste for Strawberries

AMERICAN PROFILES

A Taste for Strawberries: The Independent Journey of Nisei Farmer Manabi Hirasaki is the second book in the Japanese American National Museum's biography series American Profiles, which preserves and highlights the untold stories of Japanese Americans who have made significant contributions to the cultural, political, social, and economic life of the United States. Through the lives of the famous and the not-so-famous, this series provides a significant contribution to the literature, exploring the Japanese American experience as an integral part of American heritage.

The Japanese American National Museum is the first museum in the United States dedicated to promoting understanding and appreciation of America's ethnic and cultural diversity by preserving, interpreting, and sharing the experiences of Japanese Americans. Through the building of a comprehensive collection of Japanese American objects, images, and documents, and with a multifaceted program of exhibitions, educational programs, films, and publications, the museum tells the story of Japanese Americans around the country to a national and international audience.

A Taste for Strawberries:

The Independent Journey of Nisei Farmer Manabi Hirasaki

Manabi
& Susmi

By
Manabi Hirasaki
with Naomi Hirahara

Japanese American National Museum
Los Angeles, California

© 2003 by the Japanese American National Museum
All rights reserved. Published 2003
Printed in the United States of America

Library of Congress Cataloging-in-Publication Data available upon request.

Designer: Qris Yamashita
Production Assistance: Russell Oshita and Phillip Komai
Copy Editor: Lisa K. Manwill

Cover photos courtesy of the Japanese American National Museum.
Background: Fresh Driscoll strawberries, 2003. Left: Manabi Hirasaki, 1999.
Right: Manabi Farms, Oxnard, California, 1989.

To Sumi,
Marcia, and Mark
for their
sacrifices and support
during the many strawberry seasons

Contents

Foreword **xi**

Introduction **xv**

Part One: Fall
- Chapter One: Burning Creek **9**
- Chapter Two: Growing Seeds **13**
- Chapter Three: One of a Kind **17**
- Chapter Four: Cigars, Smoke, and the *Ichiban* Men **19**
- Chapter Five: Lessons **22**
- Chapter Six: Abe-*san* **25**
- Chapter Seven: Driving **27**
- Chapter Eight: Garlic King **30**
- Chapter Nine: Japan Pavilion **34**
- Chapter Ten: Mr. Rush **37**
- Chapter Eleven: Higher Learning **39**

Part Two: Winter
- Chapter One: On the Move **52**
- Chapter Two: Grand Junction **56**
- Chapter Three: Bismarck, North Dakota **59**
- Chapter Four: Enlisting **61**
- Chapter Five: Camp Shelby and the 522nd Field Artillery Battalion **64**
- Chapter Six: Maneuvers **68**

Chapter Seven: NAPLES **71**
Chapter Eight: MINEFIELDS
   AND FORWARD OBSERVERS **73**
Chapter Nine: MP IN MONTE CARLO **76**
Chapter Ten: DACHAU **78**
Chapter Eleven: COMING HOME **80**
Chapter Twelve: HIRASAKI FARMS, INC. **82**
Chapter Thirteen: ROMANCE AND PICKLES **85**
Chapter Fourteen: FOUNDING THE
   HOKUBEI MAINICHI **88**
Chapter Fifteen: OUR FARMS **91**
Chapter Sixteen: INDEPENDENCE **94**

PART THREE: SPRING
   Chapter One: NEW START **101**
   Chapter Two: STRAWBERRY DEAL **103**
   Chapter Three: THE CRASH OF 1957 **106**
   Chapter Four: THE BANNER, SHASTA,
      AND LASSEN **108**
   Chapter Five: NED DRISCOLL AND THE Z5-A **111**
   Chapter Six: WATSONVILLE **113**
   Chapter Seven: *BRACEROS* **116**
   Chapter Eight: PROTECTING THE ROOTS **118**
   Chapter Nine: BROKEN BOTTLES **120**
   Chapter Ten: WILD GAME **122**
   Chapter Eleven: MOTHERS, DAUGHTERS, FATHERS,
      AND PREGNANT PLANTS **124**
   Chapter Twelve: CULTURAL PRACTICES **127**
   Chapter Thirteen: HEADING SOUTH **130**

PART FOUR: SUMMER
   Chapter One: EARLY SIGN **141**
   Chapter Two: OXNARD **143**
   Chapter Three: E.F. DRISCOLL
      FARMING TRUST **147**

Chapter Four: DRISCOLL
  STRAWBERRY ASSOCIATES **149**
Chapter Five: REMOVING LEMONS **152**
Chapter Six: SHARECROPPING
  VS. SUBCONTRACTING **155**
Chapter Seven: PITCHING NETS **157**
Chapter Eight: NEW COMMUNITY **160**
Chapter Nine: RIGHT VARIETY **162**
Chapter Ten: UNION AND PROTESTS **164**
Chapter Eleven: JAPAN MARKET **167**
Chapter Twelve: ENTERING THE CIRCLE **170**
Chapter Thirteen: MANABI FARMS
  AND LONG-STEM STRAWBERRIES **172**
Chapter Fourteen: LONGEST PARTNERSHIP,
  NEW PASSIONS **176**
Chapter Fifteen: FATHER AND SON **179**
Chapter Sixteen: STRAWBERRY DIPLOMACY **181**

CHRONOLOGY **185**

ACKNOWLEDGMENTS **193**

SOURCES **195**

Foreword

By Senator Daniel K. Inouye

Japanese Americans have been breaking new ground in the United States since their immigration to this country in the late nineteenth century. Those early immigrants, the Issei, faced tremendous hardships in a foreign and often hostile land. Despite the great uncertainty that lay ahead, they knew that they were working toward a better future, not only for themselves, but for their children. Manabi Hirasaki was one of these second-generation individuals.

The Nisei generation, including Manabi, as he is fondly known by all, will forever hold a special place in American history. Living between Issei parents—whose upbringing, ethos, and citizenship connected them to Japan—and the United States—a land of individual freedoms and opportunities—the Nisei had both the burden and benefit of learning life's lessons from their immigrant parents. Manabi's earliest memories are of the garlic farm his father, Kiyoshi Hirasaki, owned in Gilroy, California, and the risks and rewards of a life in agriculture. Doubtless, such memories impressed upon a young boy's mind do not fade quickly.

As he establishes a family of his own and leaves his father's operation in 1950, Manabi's "independent journey" begins in earnest, first in celery and then in strawberries. You and I might not think about the backbreaking, intensive labor that it took in the 1950s—and still takes today—to get fresh strawberries from the fields to our kitchen tables, but Manabi Hirasaki understands. We are a mosaic of our lived experiences, both the struggles and the successes, and only by learning as we move through life will our children have hope for a better tomorrow.

When I reflect upon my own childhood in the former Territory of Hawai'i, I am keenly aware of how those early days helped to inform the decisions that I would make later in life. The fields thick with sugarcane and golden pineapple and the radiance of the land and its people will forever occupy a place in my heart. When this very land—my native country—was attacked by Japan on December 7, 1941, I immediately aided my fellow Americans to the best of my ability amidst the smoke and wreckage resulting from this attack. When the opportunity arose in 1943, I volunteered for the 442nd Regimental Combat Team and proudly served my country in the battlefields of Europe during World War II. That same year, Manabi Hirasaki also answered this patriotic call to duty and was inducted into the 522nd Field Artillery Battalion, C Battery. His basic training at Camp Shelby, Mississippi, brought him into direct contact (and sometimes conflict) with Japanese Americans from both Hawai'i and throughout the mainland, not to mention the diverse lot of people he would later encounter while serving in Europe. But despite the difficulties of those turbulent years, Manabi's pragmatic view of the war and of life embodied the type of man that he was and still is to this day. He writes, "I figured that I was doing my share for this country. We weren't thinking about ourselves, but the generations to come. I was lucky because I didn't have to sacrifice my life." We who are fortunate enough to know Manabi today and share in his wisdom and generosity are the lucky ones.

As you read about the life of Manabi Hirasaki, from his modest beginnings in Gilroy to his seat on the board of directors of Driscoll Strawberry Associates, the largest commercial distributor of strawberries in the world, consider how his life story touches your own history. As a fellow Nisei and veteran of the United States Army, I recognize the immeasurable sacrifices, struggles, and obstacles that Manabi faced. As a United States Senator who has proudly represented the State of Hawai'i in Congress since 1959, I commend his philanthropic work both within the Japanese American community and for the greater public good. As an American, I have the utmost respect for Manabi, his independent spirit, his judicious yet generous ideals, and his unrelenting integrity. He is a humble farmer at heart, a matter-of-fact man who at the same time possesses both vision and principle—a man whose examples in life we might all follow.

I want to thank Manabi Hirasaki for sharing his personal journey with us, and I also thank the Japanese American National Museum for bringing that journey to light. By experiencing one man's toils and triumphs in the world of strawberries, we can each look to the future with the knowledge that great opportunity sometimes lies in the smallest of things.

Introduction

"Tell me a story."

In the darkness of a humid Midwestern night, a group of writers gathers. My friend Essie, raised in a border town in south Texas, has just completed tales of cottonwood trees, old relatives with glass eyes, and the strength of her father, a school principal unwilling to compromise his values for expediency.

But now it was my turn. So Essie asked again, *"Tell me a story."*

Within the community of writers, this is not an unusual request. At the drop of a hat, we are to perform for our colleagues, recite poetry from memory and maybe even sing. With the aid of a karaoke machine, I can probably oblige. But under the night stars, I had to come up with something.

But I could not mine my family historic archive—not because the stories aren't there, but because they are often in pieces, shattered and sharp in some places. As a writer, I've attempted to arrange them to create a whole picture, at times getting nicked and a little wounded. To speak of these stories in public, to hear my voice take hold of them, would be too much. So I had to go back to my childhood, when my mother, a *shin* Issei, or postwar Japanese immigrant, bought me books of Japanese fairytales where I read of the tongue-cut sparrow and the crane who sacrificed her feathers to make beautiful cloth for her human husband. These were not necessarily my stories, but fairytales of a life left behind in Japan. I had no real stories to tell of a Japanese American in California, no tales of tall trees and the strong men in our community.

I've always thought it was too bad that my family doesn't have more of an oral tradition. In Hawai'i, the culture of "talk story" seems strong among the island's multicultural groups. But here on the mainland, we don't seem to tell that many complete stories around dinner tables. They seem to come out in fits and starts, with one person maybe starting something, and then another correcting or adding until the ending fizzles out, with another story beginning again.

As I've entered the business of writing biographies and organizational histories, I've had to search for the stories in people's lives. Many times it requires listening to things that are not said as much as the things that are. A lot is revealed between the lines. It's my job to figure out what these unspoken things are and to find the "knot," or the central theme of the story. That brings us to this one, *A Taste for Strawberries: The Independent Journey of Nisei Farmer Manabi Hirasaki*. The son of a Gilroy farmer who reconstructed a Japanese showcase house in the middle of garlic fields, Manabi entered strawberry growing in his late twenties. He rose to become the first non-European American board member of Driscoll Strawberry Associates, the largest strawberry distributor in the world. More than a study of strawberry farming, this is a story of one Nisei's attempt to "enter the circle" of a corporation. Like certain ambitious third-generation Japanese Americans in corporate and academic America, Manabi has sought to be a decision-maker, an owner of a piece of the pie. Rather than relying on manipulation or grandstanding, he did this with his easygoing personality, connections through his father, and attention to details.

Manabi is more than a farmer who fought to enter the circle; he also is a philanthropist. He has given much to the Japanese American National Museum, as well as to the Go for Broke Educational Foundation and other nonprofit organizations. Representing his enthusiasm for education, Manabi's proudest project to date is the Japanese American National Museum's Hirasaki National Resource

Center. He has funded films and books. His name means "to learn." It is interesting that after so many years of working with those who are not Japanese American, the Hirasakis have decided to plant themselves squarely in Japanese America. It is as if an old friend, perhaps a spirit left by his father, Kiyoshi, and mother, Haruye, has harkened Manabi and his wife, Sumi, back into the fold.

I can't remember when I first formally met Manabi Hirasaki, but I remember feeling his presence. There was a softness, a gentleness in his manner. An easy laugher, yet a straight shooter—no false compliments or underlying political agendas. Alongside the humility of a farmer, there's also the stylish side, most evident in his choice of understated yet elegant shoes. Topping his traveling outfit is usually a fancy fisherman's hat, often adorned with a Japanese American National Museum logo pin.

Of course, always accompanying Manabi is Sumi. One of the Japanese American National Museum's early administrators, Nancy Araki, told me the story of the Historic Building's scheduled opening in 1992, the day after the Los Angeles riots happened to break out. All the staff members and volunteers had to lend a hand in revising the opening ceremony, which was moved from outdoors to indoors. In the early morning hours, before the ceremony began, Sumi was there with a broom in her hand.

In terms of models of published autobiographies and memoirs penned by Nisei men, there are several examples. Senator Daniel K. Inouye's *Journey to Washington* (1967) and Mike Masaoka's *They Call Me Moses Masaoka: An American Saga* (1987) discuss their political worlds, while Karl Yoneda's *Ganbatte: Sixty-Year Struggle of a Kibei Worker* (1983) examines the life of a labor activist. Daniel I. Okimoto's *American in Disguise* (1971), Bill Hosokawa's *Out of the Frying Pan: Reflections of a Japanese American* (1998), Jim Yoshida's *The Two Worlds of Jim Yoshida* (1972), and Minoru Kiyota's *Beyond Loyalty: The Story of a Kibei* (1997) delve into the Nisei identity in light of World War II and its aftermath. George Takei's *To the Stars: The Autobiography of George Takei, Star Trek's Mr. Sulu* (1994) also touches upon the legacy of the wartime incarceration of Japanese Americans, but in the context of a Nisei who became a television star.

As writer Brian Niiya has pointed out to me, many of these autobiographies, including Manabi Hirasaki's story, share common traits. Many discuss at length their professional journeys, yet not so much of their personal lives, specifically their spouses and romantic partners, children, and siblings. This directly contrasts with Sansei David Mura's memoir *Where the Body Meets Memory: An Odyssey of Race, Sexuality, and Identity* (1996), a raw exploration of a relationship and deceptions between a husband and wife, a Sansei man and Nisei parents. The Nisei memoirists also document their own parents, the Issei, but in a more reverential manner. In these autobiographies, World War II is invariably a pivotal time, when men of this generation had to decide whether or not to fight on battlefields or within detention camps.

My personal observation is that Nisei men entering adulthood in the 1940s and 1950s have had to negotiate their way in an outside world that was at times overtly hostile to Japanese Americans. As a result, their focus has been on their relations with the "outside," rather than their inner domestic lives. Making their mark in their professions has become a measuring stick for their success as minority men. Relationships with their spouses, although important, are therefore usually not verbalized. One exception is a self-published work, Dick Kobashigawa's lovely *Hitome-Bore: A Kibei-Nisei Story* (2001), which is an ode to his late wife. (*Hitome-Bore* means "love at first sight.") As a Kibei Okinawan American gardener, artist, and poet, Kobashigawa seems more emotionally expressive than the Nisei men who have made their livelihood in business or politics. Or perhaps it is indeed easier to articulate these feelings once the loved one is gone.

There are ways that Manabi's story is different from other Nisei autobiographies. His family "voluntarily" relocated to Grand Junction, Colorado, from the West Coast in March 1942, so he did not experience life in an assembly center or concentration camp. He decided to leave his family's successful farm business in 1950, a move which effectively led to its closure. He did not learn strawberry growing by toiling in the fields with his family as a sharecropper, or subcontractor, as he likes to describe it, but was lucky enough to enter the business in a matter of days due to the lead of a European American attorney friend. It is also notable how he empathizes when people, including his former field laborers, experience economic or racial hardship. Because of these Nisei sensibilities of patience and humility, Manabi has, in essence, tilled his life for future seasons of good crops.

Through numerous interviews, Manabi verbally related to me the material for these chapters. As the project continued, I began to clearly hear a rhythm in Manabi's speech. This voice—simple, self-deprecating, and sometimes humorous—opened doors to the world of a Depression-era boy hunting for rabbits with his baseball bat, a young man seeking to avoid minefields while lining communication wire for his military unit, a thirtysomething man staking out new land in a place by the Pacific Ocean called Oxnard. Phrases like "the strawberry deal," "my parents didn't preach to me much," and "jumping from place to place" made his personal life history come alive. Even as I read the words on the page from the transcriptions provided by the Japanese American National Museum, I could feel Manabi's presence. This voice needed to be retained in the form of a first-person narrative. To supplement Manabi's stories, interviews with his friends and colleagues were then excerpted throughout the book.

A Taste for Strawberries does not intend to cover the comprehensive history of the Japanese American involvement in strawberry growing. As its title reflects, this memoir should explain why Manabi has a taste *for* strawberries, rather than how a taste *of* strawberries has shaped our nation's agriculture and ethnic communities.

Among the sources that would be helpful for those doing research on strawberries and Japanese Americans are Lane Hirabayashi's articles "The Delectable Berry: Japanese American Contributions to the Development of the Strawberry Industry on the West Coast" (Japanese American National Museum, 1989) and "The Issei Community in Moneta and the Gardena Valley, 1900–1920" (*Southern California Quarterly* 70, Summer 1988), co-written with George Tanaka. The latter article delves into the productive strawberry growing farms operated in Southern California, and most particularly Gardena, a topic that is not addressed in this book. For strawberry growing in Northern California, refer to Miriam J. Wells's comprehensive work *Strawberry Fields: Politics, Class, and Work in California Agriculture* (Cornell University Press, 1996). The National Museum itself has a wonderful collection of oral histories conducted by Vernon Takeshita and Nancy Araki. Other oral histories of pioneering strawberry growers, such as Unosuke Shikuma, can be found in libraries in Watsonville and at the University of California at Santa Cruz.

Kazuko Nakane has written a beautiful book on the Pajaro Valley area, *Nothing Left in My Hands: An Early Japanese American Community in California's Pajaro Valley* (White Pine Press, 1985). For more information about that region, I would highly recommend that work, in addition to Sandy Lydon's *The Japanese in the Monterey Bay Region: A Brief History* (Capitola Book Company, 1997), which includes some sections on Watsonville. Both the Pajaro Valley Historical Association and the Gilroy Historical Museum also offer other valuable materials. For information on Oxnard, refer to Yoshio Fukuyama's "The Japanese in Oxnard, California, 1898–1945" (*Ventura County Historical Society Quarterly* 39, 1994).

There are times throughout the book that things are left unsaid with minimal elaboration. As in life, we may have to read in between the lines. We have decided not to embellish the spare prose in order to maintain the Nisei voice of Manabi Hirasaki. He is speaking to us, and in listening we will be fully involved in the experience of piecing his life together. This book is only one of hopefully many stories that will be shared around dinner tables across the United States. My personal desire is that we Americans of various ethnic cultures continue to collect anecdotal snapshots—however brief or simple—of our families and community histories. So the next time someone asks us for a story, we will indeed be able to tell one.

NAOMI HIRAHARA
July 2003

PART ONE: **FALL**

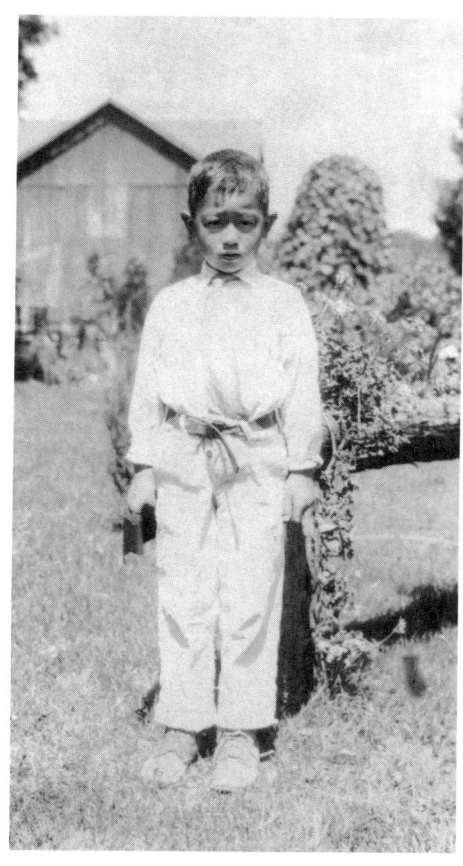

A young Manabi Hirasaki in his hometown of Gilroy, California, 1929.

Hirasaki family, 1938. Gift of Manabi Hirasaki, Japanese American National Museum (95.77.1).

TOP: *Manabi Hirasaki knifing a field of garlic in Gilroy, California, 1939.*
BOTTOM: *Manabi Hirasaki inside his father's home, partially reconstructed from the 1940 World's Fair Japan Pavilion. Gilroy, California, 1941.*

TOP: *Manabi Hirasaki enjoying the pool at Gilroy Hot Springs, ca. 1938.*
BOTTOM: *Manabi Hirasaki (left) and Bill Kuwada (right) at Gilroy Hot Springs, ca. 1938. Photograph by Harry Tanaka.*

Manabi Hirasaki vacationing on the North Shore of Lake Tahoe, ca. 1946.

Manabi Hirasaki's Gilroy High School senior portrait, June 1941.

CHAPTER

BURNING CREEK

My father took a lot of chances. He bought a four-hundred-acre ranch in Gilroy, California, in 1930 at the height of the Depression. To start, he borrowed five thousand dollars from a friend as a down payment. I knew the hard time he was having, but he never said anything. In 1930 he was only thirty years old.

My father, Kiyoshi Hirasaki, was born on March 1, 1900, in a town called Kagami in Yatsushiro, located in Kumamoto, Japan. He was the second son. His younger brother, the third son, stayed back in Japan and had six boys. Two of them are in Kumamoto operating a wholesale business in which they make flat and unfinished *tatami* mats called *goza* in Japan. A lot of the farming around there was and is still centered on growing straw for *goza*. Even my mother's family, who was from the neighboring city, Miyahara, was involved in that business. Miyahara back then was a small town of about one hundred people. The straw was woven together with a needle and thread, and then the loose mats were stuffed with hay and turned into *tatami* mats. These mats were then loaded onto trains and delivered to towns like Tokyo and Yokohama.

In the early days, even though our family had farmland in Kumamoto, it wasn't worth much. There was no work. And many young men were being conscripted into the military. That's why in 1911 my grandfather Tokutaro took the Number One son, Tokuzo, to America. After he was settled, my grandfather went for my father in 1914, when he was only fourteen years old.

They first went to San Jose, located in the Santa Clara Valley, just south of San Francisco. The Valley includes cities like Milpitas, Mountain View, Cupertino, Saratoga, Los Gatos, and my hometown, Gilroy. Japanese first started farming as migrant laborers in the valley in the mid-1890s. They picked strawberries in late spring, then worked the apricot, pear, and prune harvests in the summer.

The Mayedas, another family from Kumamoto, had also settled in San Jose. They were from the same village as my mother, Haruye, and had a small boardinghouse in San Jose's Japantown, or J-town. It was upstairs on Sixth Street next to a Chinese restaurant and a barbershop. The daughter, Matsue Kami, explained what it was like in San Jose in the 1910s:

> *A long time ago, many went to work as* buranke katsugi *(those carrying blankets). During the summer they had work, but during the wintertime it rained, so they came to stay at the boardinghouse. That was the case of [Manabi's] grandfather. They carried a blanket like a sleeping bag, going from work to work. During the summer some would come back and forth; sometimes owners would come on horse and buggy to take workers to the field, and then they would take them back to the boardinghouse. Just like the cowboy movies.*

My father stayed at Mayeda boardinghouse while he was a schoolboy. He was a teenager while Matsue was about five or six years old. Since my father didn't have his family nearby, Matsue hung around him like a little sister.

Maybe because of this type of life, my father never worried about how he was going to make a living. When my father bought his ranch, I knew he was having a hard time. My mother would ask my dad to buy a loaf of bread for our lunch the next day, and then he'd forget it. At least he said he forgot it. But I don't think he had the money to buy a loaf of bread. And then there was a time when the bakeries would travel to all the ranches to deliver bread and pastries. I remember that the bakery man would ask my mother to see if my father could pay off some of his bill.

My dad didn't believe in banks and they didn't treat Japanese well anyhow. Instead of saving money, he would just spend more to progress. Whatever he got he'd just plunge back into the business again, so he wasn't the kind of person who had extra money. He'd just spend it buying equipment and making other improvements. In that sense he was a real Kumamoto man.

Back then, they didn't give the Japanese the chance to buy the good, established ranches. You had to pretty near develop your own. My father's land wasn't developed; it had been a dairy farm and, as a result, was just dirty. It was full of willow trees and all kinds of other shrubbery. In the wintertime, Llagas Creek, located below the ranch, would flood, creating a wet swampland.

My father's ranch was around four hundred acres, and willow trees grew on about one hundred acres of it. So he cut down all those willow trees. Old-timers in Gilroy still talk about what he did next.

The willows were all chopped down and piled, ready to burn. One foggy day in the fall, my dad poured gasoline over the stumps and limbs, torched the pile, and then took off to San Jose, which is about thirty miles away. It was the biggest fire they had ever seen in Gilroy. It was right in the middle of the ranch, so the fire wasn't going to go anyplace, and there weren't any environmental people back then. In those days, smoke was smoke. That's all.

That's the kind of thing my dad would do. He wouldn't ask for permits or anything like that. He built a fire once more. At that time, Llagas Creek was completely dry. He filled it up with stumps and he lit it. He then had a big squabble about the fire with his neighbor. The sheriff came out and told my dad to take the remaining stumps out of the ditches. But that night, he torched them again. Afterward, he would say, "I don't know what happened."

One time my dad rented the dairy barn to some wine bootleggers. The ranch was still pretty new; he was about thirty-three years old. After the bootleggers started bottling, the sheriff's deputies came on a raid. They went through the whole wine operation. They broke all the tanks and smashed the wooden barrels. Then they began looking for my dad.

In those days, most of the laborers were Filipino. The deputies went around asking the workers, "Where's the boss?"

"Boss not around," they answered. "He's gone someplace, I guess."

But my dad was sitting right there in one of the cars. He was a young-looking man, so he wasn't recognized. I saw him there myself, and the workers all knew that he was hiding, but the deputies never knew it. They never came back for him, and I don't know whether they ever caught the bootleggers.

CHAPTER

Growing Seeds

Many people have heard the phrase "planting seeds," but my father was in the business of "growing seeds."

Another Issei who grew seeds was the father of my childhood friend Hiromi Nagareda. Yoshio Nagareda first came to California as a schoolboy in 1914. Later he returned to Hiroshima to get married and, together with his wife, he traveled to Gilroy one month before Hiromi was born on July 31, 1922. Yoshio grew seeds for Pieters-Wheeler Seed Company. The owner, Lin Walker Wheeler, was originally from New York but grew up in Wisconsin. He spent most of his youth prospecting and working on the railroad before getting into seeds. He eventually bought controlling interest of the Pieters-Wheeler Seed Company around 1911.

Wheeler didn't farm himself, although the company had some trial ground for experimenting. He instead had most of his seed crops raised by Japanese growers. The company leased the land and families like the Nagaredas would grow the company's seed crop, as well as a side crop of garlic, tomatoes, or something else on their own.

The Nagaredas grew lettuce, mustard, and onion seed—all specialties in California. In fact, Santa Clara Valley was known for its variety of seeds. There's something about the combination of coastal air and the coastal mountain range, which traps the warmer and drier air of the interior. These two things made for good seed growing conditions.

Seed crops don't look like regular vegetable crops. Stalks stand several feet from the ground. In the case of onions, the bulb is up on the top. To produce lettuce seeds, you need to slash the head of a lettuce with a knife; those shoots may then grow up to reach the shoulder of a man. Depending on the type of crop, they usually grow the seed in spring and then the seed matures by July or some months after that.

My friend Hiromi once explained how they separate the seeds from the plant:

> *In the old days, the field workers, usually Filipino men, used to get a Japanese* kama, *a hand sickle. They would cut [the plant] with the* kama *and carry it on their backs to a big canvas, about the size of a 50 x 50 room. There they rolled it with a roller, either flat or round, that was pulled by a horse. The horse would go around in circles, and later we [rolled the plant with] a tractor. I used to do a lot of that before the mechanical threshers came in.*

My father worked for Kimberlin Seed Company, a small business in the Milpitas area of Santa Clara. That area is now the city of Fremont, where an automobile plant produces both Toyotas and General Motors brands.

The most critical thing in seeds is to not mix them with any other varieties. They have to be pure. If you mix them, then you can't sell the seeds. Take lettuce, for example. People want a certain kind of lettuce; it has to be that one, not any other kind of variety. You also have to be careful of getting the best seed. The company sends out inspectors to check the germination percentage, planting the seeds and then examining them. The product needs to register at least 90 plus on a scale of 100 or else be faced with a low-quality classification for pricing.

Another Gilroy friend, Masaru Kunimura, who everyone calls Moose, is also familiar with the seed business. His father, Kazuto, worked for Wheeler's competitor, Rohnart Seed Company, located in nearby Hollister. Moose explains:

> *Nobody knows what the quality of that one seed is going to give you. But usually everybody is trying to get the biggest pumpkin, so they naturally take the seed out of the biggest one.*

When my dad was with Kimberlin, he used to grow a lot of onion, carrot, and anise seed. The anise, which gives licorice its flavor, was used for pickling, and San Jose was the cannery capital of the nation at one time. When I was young, San Jose had about twenty-two canneries—although none of them were Japanese-owned. There was big agriculture in the area, and San Jose had good production. In general, agriculture businesses in California had long production seasons. Back East, when the harvest was over, canneries would be finished. But in California these canneries could do two or three different crops during spring, summer, and fall. When I was about twelve or fifteen, someplace in there, I saw the same companies handle peaches, tomatoes, and even pickles. Kimberlin had all different kinds of pickles; my favorite were those small, sweet ones. They were called lady fingers. The cannery would have assorted barrels of pickles set aside for their own people for lunch. Every time I'd go into a pickle cannery, I knew where those barrels were, so I ate my fill of them.

In 1931, when I was eight years old, my dad had a small seed shop for both retail and wholesale trade on Jackson Street in San Jose's Japantown. Before that, I had only lived in Gilroy—on Frazier Lake Road and then a second place we rented from a Mrs. Brownell. When we moved to San Jose, there were about four of us children by then and we lived in the back of the store, which was called Hirasaki Seed. All the seed was in sacks stacked on one side. My mother weighed out the seeds and put them in little packages that were sold retail to farmers.

I played in the store and outside as well. Sometimes I even went into town to see old cowboy movies starring Buck Jones, Hoot Gibson, and Ken Maynard for ten cents a picture. My friend's mother was a nurse at the Kuwabara Hospital in town, so we used to play in the front yard there all the time because there were a lot of trees and shade. Next to the hospital was the home of Kunisaki "Kay" and Kane Mineta. The year I lived in San Jose was the year their son Norman Mineta was born. Mineta went on to become a San Jose city councilman and mayor, and then later a U.S. congressman and the first Asian American on the cabinet.

Those days were hard on my mother. She had a lot of mouths to feed. One day she couldn't get a sack of rice because my dad was still out in Gilroy farming. She had to have our landlord, Mr. Takeda, okay a credit for a sack of rice from Dobashi Market, which used to be called Kinokuniya. That market is still there and is run by the grandchildren and great-grandchildren of the original owners.

After a year, we moved from San Jose to the newly purchased ranch in Gilroy. My dad was still growing seeds, but this time he turned the abandoned cheese plant on the ranch into a seed cleaning mill. He would run the seed through the mill and separate the seed from the flowers, weeds, and trash. I can still remember seeing the carrot seed fields in bloom; the flowers were white. But in a few days the bloom went away, and all that was left was this beautiful color, the color of golden corn husks.

CHAPTER

ONE OF A KIND

When I was born in San Jose, my mother named me Manabu. Manabu means "to learn," and I guess she got that word from the front of a schoolbook. It later got misread and misspelled on some papers as "Manabi" and it stayed that way. So I say that I'm one of a kind.

Names are kind of funny. Even though my dad was named Kiyoshi, everyone called him Jimmy. A lot of Japanese names get anglicized because they are too hard to pronounce. Other than the time I served in the U.S. Army, I never took on a so-called American name. I figured that people would just have to deal with my Japanese name.

After me came Mineko, Fumiko, Michiko, Aiko, Hisashi, Shinobu, and Midori. I was the oldest child and then came four girls. That meant I kept more to myself. I had my own room. When the fourth sister, Aiko, was born in our house with a midwife, all the girls were together in another room. They were all excited when they heard the baby's first cry. By then I was ten years old and, like other boys my age, I wasn't interested in babies.

My Number Three sister, Michiko, remembers walking with me after the school bus dropped us off at the highway, one mile from their house. I don't recall this particular story, but she says that I made her carry my books. According to her, she told me one day, "No, I'm not going to carry them." Then I answered, "You better carry them or I'll leave them here and they'll be there all night." When I'd walked away, behind me was Michiko, carrying my books.

My mother was soft and quiet. She had many friends. One of them was Mrs. Kami, from her hometown in Kumamoto. Like my father's family, my mother's family, the Yonemitsu, was involved in the growing of hay for *goza*. When Mrs. Kami's first husband, an asparagus farmer in Sacramento, died in 1937, she came to live with us in Gilroy for a short time. They would knit and play the piano together. They sang childhood songs like "Hata Popo." Sometimes she would tell Mrs. Kami that she was lonely and missed her sister.

My mother was religious in a way—there was certainly a little religion in her family. But with so many children, she didn't have time to go to church. For Gilroy people, that meant traveling all the way to San Jose for the large Buddhist Betsuin temple in J-town. We didn't have our own church in Gilroy—unless you count the makeshift Koyasan church, a Buddhist sect, that we had on the other side of our four-hundred-acre ranch. The church was actually a piece of white canvas stretched over a wooden frame to form a tent. A traveling Koyasan priest, somehow connected to my mother's family, would come into town from Portland, Oregon, about once a month, or at least once every two or three months, and this would be the place where he would meet with a few people, conduct services, and say a few chants. He used to be called Reverend Kimura, but he later changed his name to Reverend Henyoji.

I've met some younger generations of Henyojis since then. When they introduce themselves, I ask them, "Are you sure your name is Henyoji? It's not Kimura?"

And every time, they look shocked. "How do you know?" they ask.

There's just something funny about names.

CHAPTER

CIGARS, SMOKE,

AND THE *ICHIBAN* MEN

According to my friend Moose, my father and another grower, Mr. Morita, were the *ichiban* men of prewar Gilroy. *Ichiban* means "Number One" in Japanese, and Moose knew who was who because he hung around his uncle Junichi Tanaka's grocery store on Monterey Street, where the *ichiban* men apparently gathered and played a Japanese card game called *hana* after a long day in the fields.

In the 1930s this stretch of Monterey Street was mostly comprised of Chinese and Japanese stores. The Chinese actually had come first, decades before, when Gilroy was known as the tobacco capital of the United States. According to community historian Patricia Baldwin Escamilla, nine hundred Chinese men had been recruited to roll cigars at a factory owned by the Consolidated Tobacco Company of Gilroy—the largest of its kind in the world. When the government outlawed Chinese immigrants from entering the country, the business closed down. Dairy, cheese, and prunes then became Gilroy's signature products.

Moose remembers relations between the Chinese and the Japanese as pretty neighborly. Both had dealt with discrimination and, for that reason, they stuck together, at least on Monterey Street in between Seventh and Ninth Streets. The Japanese businesses included Tanaka Grocery, Horita Laundry, Koga Restaurant, Nakamura Soda Fountain and Pool Hall, and Gilroy Shokai, a grocery store owned by the Fujimoto family. The rest of the block was made up of Chinese restaurants, curio shops, and gambling dens. Like other towns of that time, Gilroy also had its share of *joro*, or brothels.

Everyone was familiar with me so whenever I went into town with my dad I freely went in and out of the stores and even the gambling dens. I remember seeing men playing a lottery like modern-day keno, writing numbers down and wagering before selected numbers were called out.

As I mentioned earlier, the Chinese and Japanese got along pretty well—until the conflict in China broke out in the 1930s. Then Issei started to send money and other supplies to Japan to help in the war effort. They even collected aluminum foil from gum wrappers and tobacco, balling it up and shipping it to Japan for the aluminum supply. They were first-generation Japanese, and their Chinese neighbors were first-generation as well, so it makes sense that there would have been hard feelings even though the war had little to do with the United States.

Because my dad had come over when he was young, he was one of the few Issei who could speak English. He was friendly with English speakers as well as other Issei. He would even set aside ten to twenty acres of a crop for the Japanese community. "You Japanese come and harvest it and you can have the profit," he told the community leaders. Everybody came out; it was more like play than work. Mr. Morita would do the same thing as well.

Moose remembers that the community held Japanese movie screenings from time to time. The movies were silent, and a *benshi*, a type of actor, was brought in to provide the dialogue. The names of the financial sponsors and the amount they gave were listed beside the stage on wooden plaques or paper signs, arranged with the largest donors on top. Moose recalls:

> *Mr. Hirasaki and Mr. Morita, they're number one and number two. Nobody surpassed what they put up...If Mr. Hirasaki put up X amount of dollars, [and] Morita, X number, they're*

never the same. The two names go right up. After that, everybody's name goes up. Nobody, nobody goes above that.

Later, after the war, when my father became involved in starting a new Japanese-language daily newspaper in San Francisco, *Hokubei Mainichi*, as well as supporting the Buddhist Churches of America (BCA), I'd criticize him for doing too much. I was still young then, and I didn't understand his legacy as an *ichiban* man.

CHAPTER

LESSONS

Every day after public school, all the Japanese kids would go to the Japanese language school, *Nihongo Gakko*. Regular school was out at 3:30 p.m. and you had to be at Japanese school by 4:00 p.m. When I spent a year in San Jose, there was a little bit of Japanese teaching after Buddhist class on Sundays, but I didn't pay much attention to it. I didn't care much about learning Japanese, and you *had* to care or else you wouldn't pick the language up. I just didn't have that kind of brain. And even the brains in the class eventually forgot what they learned, too.

In terms of public school, I first went to a country school out in the Old Gilroy District on Pacheco Pass, about five miles from town. Later, when I returned from San Jose, I attended Elliot School, more in the city. Some country schools were a bit different, made up of one room with multiple grade levels. Farm children without transportation went to these schools. Since many Japanese farm families moved from one piece of land to another, their children ended up going to several different schools over the years.

My mother's sister was married to a Buddhist priest, so my mother's family was more into learning than my father's. Education was always important to my mother. In middle and high school my favorite subjects were geography, history, and math. English, especially the grammar part, was harder for me. In high school I majored in agriculture. My agriculture teacher, Forest Rycraft, was grooming me to get a Bank of America achievement award in farming. To write different reports, I had to ask my dad a lot of questions; normally I hardly spoke to him because he was so busy on the ranch. But it was interesting to learn from him when I could. He was already following the practice of a modern farmer: he rotated the crops on his 1,600-acre ranch. He had also created a drainage system of canals so the winter rainwater that accumulated in Llagas Creek would not flood the land. I understood what he was after, and he would work with me to complete my papers.

During this time, I also played basketball. Gilroy High School was small—only about five hundred students in all—so it wasn't hard to get on a sports team. Earlier, I played lightweight football but quit before my senior year because I couldn't compete with the larger boys. In terms of basketball, we were just a fair team; we'd win some and lose some. The larger schools in Santa Clara, Palo Alto, and Mountain View were much better. I always say that hungry people get ahead—in terms of the guy who works hard all his life to prove something. Before the war, it was the Italians, like Hank Lusetti, the star basketball player at Stanford University, or else Eastern Europeans. After the war, it was the African Americans.

The Nisei boys also had our own Japanese community basketball team, the Gilroy Eagles, in the 4 County Athletic Association, also known as 4C2A, which included Salinas, Monterey, Santa Clara, and Santa Cruz Counties. We didn't have a real coach and we didn't have a gym to practice in. That made it pretty hard to be really good. But it was fun going to these different towns and getting together for skating parties and dances after our games.

My parents themselves didn't preach to me much. They did teach me to organize myself and to be on schedule. It was also important for us kids to be respectful of other people. As a result, we never had people waiting for us—in spite of being such a large family.

In terms of discrimination, I can't recall too many incidents, maybe because Gilroy was such a small town. I particularly knew the whites, the *Hakujin*, longer than most of the Japanese because we were in grammar school together. I do remember something about going to the movie theater, though. When I went by myself, nobody would tell me where I could sit. But when I was with our Filipino workers, we always sat on the side. So although I didn't feel discriminated against, there were times when I was with people who were.

CHAPTER

ABE-*SAN*

After my father bought the Gilroy ranch, we had workers come out and live there. Since it was still the Depression, we didn't have enough money to pay everyone, so we provided room and board instead. My mother ended up having to cook for that whole head crew, which at times numbered fourteen, fifteen, and even twenty men. The harvest crew, on the other hand, fended for themselves.

My mother didn't have a maid but she did have an old bachelor that helped around the ranch. We called him Abe-*san*. He was a short Japanese man with a round face. He was heavy; we kids always said that he ate too much. He didn't say much to me. In fact, it seemed like he was grunting all the time. He understood what we said to him, but I don't think he spoke a word of English.

Our house sat in the middle of the ranch, but we also had another building for the workers.

Abe-*san* lived there, right next to the kitchen, and he would start all the fires for my mother, both for the stove and also the *furo*, or bath, for the workers. He would bring in the wood and keep the fires going. My mother would use a big *nabe*, a Chinese wok, to fry foods and cook *okazu*, little bits of pork and chicken. She used another wok to steam rice.

I was too young to cut wood, so during the wintertime I went out in the back and picked tree mushrooms that grew on willow trees in the swamps. We also raised chickens and had about four or five hogs. The Filipino workers would slaughter the hogs for us because they wanted the blood for a sauce, their delicacy. When I was about ten years old, I watched them kill a hog. The workers first slit the hog's throat and then drained the blood into a big pan placed underneath the animal. The hog would move around a bit, so the workers had to hold it down for a while.

When we didn't have any meat during the wintertime, I would go hunting for jackrabbits or cottontails. They were plentiful behind the ranch when it rained. The ground would be flooded and the rabbits would be stranded on islands of land. I went out there in my boots and hit them with my baseball bat; I'd get about half a dozen. Abe-*san* would skin them. Mom would cook the meat with mushroom to make *okazu*. That was very good.

Every Saturday Abe-*san* would go and buy his jug of wine for four bits, or fifty cents, a gallon. In those days, Italian families made their own wine, and he knew of a place to supply his weekly habit. Since he was a steady customer, they'd give him one glass of wine gratis. Even if someone offered to buy a glass for him, he'd say, "No, I get that for free."

We eventually went separate ways because of the war, but we met up again in San Jose. In the end, he died at a Salvation Army home for old bachelors, full of men like Abe-*san* who at one time went from door to door, looking for new opportunities.

CHAPTER

DRIVING

I began driving cars when I was about twelve, thirteen years old. I was already behind the wheel of a tractor when I was nine. I drove the tractor in the summer and helped around the ranch as much as I could. I usually did the work that didn't require a lot of precision—like towing cultivators and other farm equipment—but not at the fast speed that they're doing everything now.

Driving together to events like funerals was one thing I did with my father. He would be so tired after a long day of work and it would be too hard for my mother to get away to drive him, what with all my brothers and sisters to take care of, so my dad and I would go, and I'd do the driving. In those days, you never heard about Japanese staying in hotels and motels. If you had to rest, you slept in your car. More often than not, you'd drive all the way home through the night.

We'd go to San Jose, which was about thirty miles away. One time I fell asleep at the wheel. When I woke up, a truck was coming right at me. We both swerved and missed each other. My father never knew what happened because he was fast asleep. I even drove one hundred miles up north to Stockton. When we went to the funeral of Matsue Kami's first husband, we traveled all night. My dad could only think about starting the centrifugal water pumps for irrigation, so we had to get back by five o'clock in the morning. We had to prime the pumps by hand—nobody had electricity on water wells at that time—and then start up the gas motor. In the 1930s the water table was high so wells could be as shallow as twenty feet below ground level. Today, wells need to be drilled down at least six hundred or seven hundred feet to hit water.

Around 1936 my father bought a used 1934 Chrysler sedan and completely overhauled it. One time we actually traveled all the way to Texas in our fixed-up Chrysler. We went to visit our Uncle Tokuzo, who had married into the Kishi family of Terry, Orange County, Texas, located near the coastal town of Beaumont. The senior Kishi, Kichimatsu, started one of the Japanese rice colonies in Texas back in 1907. An educated man, Kichimatsu faced a lot of hardships, including the collapse of the rice market after World War I. New canals were opened up in the area, causing saltwater to flow into the rice fields. He later discovered oil on his land and started a petroleum company. George Hirasaki, Tokuzo's son and my cousin, is now a professor at Rice University specializing in oil recovery.

The year we went to Texas, however, times were tough. The Great Depression had hit the area hard and many of the rice fields had been converted into truck-farming crops. It was wintertime and cold traveling so many days by car. It's a miracle that we were able to make it both ways.

When I got older, I had access to the family car because my dad used his pickup most of the time. That helped my social life, although I didn't go to school socials like the prom. Although I dressed better than most teenage boys in terms of everyday wear, I didn't have a suit or necktie. Besides that—let's face it—I wasn't a good dancer. In fact, in the Gilroy High School yearbook, I stated that my ambition was "to learn to jitterbug."

I dated, but mostly out-of-town girls. To tell you the truth, the hometown girls in my class were all too smart for me. In terms of dating, it was nothing serious. Time changes fast; one month of dating the same girl is a long time. Because I had access to the family car, my friends always tried to get me to double-date with them. One time I got talked into taking my friend and two sisters we knew to see *Gone with the Wind* in San Jose. We went to the 9:00 show, and by the time the picture was over, it was past midnight. Looking back, I don't know why I went because I had already seen the movie back in Gilroy.

One of my good friends at the time was Harry Tanaka. Harry's mother helped run Gilroy Hot Springs, which was located in the hills about fourteen miles from town. I would drive up a winding road to a large pool beside some cabins constructed by the Nishiura brothers. According to historian Patricia Baldwin Escamilla, Gilroy Hot Springs goes back to the 1870s, when a man named George Roop created a resort to take advantage of the springs, which apparently flowed at 107 degrees. A farmer, Roop was also a devout Christian and insisted that no drinking take place at the hot springs. So all our parties there had to be without any alcohol. Mr. Sakata bought the springs in the 1930s.

Harry Tanaka was with me when I snuck out in my dad's 1938 De Soto sedan to go to the World's Fair in San Francisco in 1940. Coming home, I fell asleep at the wheel. The De Soto was heading toward the Bayshore Highway. A truck pulling a trailer drove across on the highway, and our car hit the tongue of the trailer before bouncing back into the street. Luckily, no one got hurt besides some bruises, but the De Soto was totaled.

The car was towed back to the ranch. My father didn't say a word. He was the kind of person who forgets everything and just moves on. Regarding the smashed-up car, he was probably thinking, "Got to buy another one."

CHAPTER

GARLIC KING

My father never put all his eggs in one basket. As I mentioned earlier, he was a farmer who constantly rotated his crops. He grew onions, chili peppers, tomatoes, sugar beets, and spinach on his 1,600-acre ranch. Five hundred acres were reserved for vegetable seeds. But by 1937 he was also known as a garlic man.

He wasn't the only Japanese farmer growing garlic. Yoshio Nagareda was one of the early garlic planters, back in the times when there wasn't that much of a demand for the crop in California. According to my friend Hiromi Nagareda, most of it, especially the number-two quality, was shipped out of the country to places like Cuba in the early 1900s.

Back in the 1890s, there were mostly orchard crops, including apples, apricots, cherries, peaches, pears, plums, and nuts. Gilroy, in fact, was known as the prune capital of the state during the 1920s and 30s. Italians had brought over row crops like tomatoes, peppers, and onions. They grew garlic too, but usually for their own use. It was the Japanese who first learned how to make money off of it. By 1938, 50 percent of the garlic grown in California was on farms operated by Japanese. If you take a look at Masakazu Iwata's book, *Planted in Good Soil*, you'll learn that in 1938, Japanese Americans farmed 1,155 acres of garlic, a crop worth close to $225,000.

To grow garlic, you need cold winter rain in dormancy and hot, dry summers to cure the bulbs. If you don't dry them well, they cannot hold their quality. You couldn't raise garlic in nearby Watsonville, for instance, because of the high humidity, but both Gilroy and neighboring Hollister were good for garlic. For the same reasons, they were also good for the production of seeds like carrot and onion. You need to have the seeds dry, and the only way to do that is out in the sun with low humidity. Just like raisins in the Central Valley.

Garlic tended to get root disease fairly easily, so the farmers needed to rotate the crops at all times. If the ground got diseased, it would be bad for future plantings because they wouldn't get the yield or the quality to be able to store the harvest. They wouldn't be able to replant on the same soil either.

Planting was done in the fall. In those days, you had to crack the garlic by hand to get the seed. Each clove was a seed, so you would plant each individual clove, placing them three to four inches apart in two rows on a bed. Back then we thought we had to plant the roots of the garlic clove facedown, but nowadays they just put the clove in the ground any which way because they found out the roots are going to naturally go down toward the moisture.

Harvesting usually took place in June and July. We would use a cutter bar, or knife, to slice the roots, pull the garlic out of the soil, and then stack it into neat piles called windrows. Leaves from the plant were wrapped over the bulb to protect it from the direct sun and the rain. (The garlic would turn black when wet.) After the garlic was cured in the sun, we'd snip off the tops and the roots, which would be dried after two to three weeks.

My dad estimated the cost of producing a pound of garlic at two to three cents; he would say, "I could make money on three cents." Just before the war, the cost went up to six cents, but he was probably selling the garlic at thirty-nine to forty cents per pound. He made a lot of money. He had about ten to twelve Japanese American subcontractors (I don't like the word "sharecroppers"), and they would farm around thirty acres a piece. My dad would pay for the supplies and recommend how to take care of the garlic, but they would do the planting and harvesting.

One of our subcontractors were the Nakayamas. The father, Fujio, came to the United States from Saga, Japan, at the age of fourteen, the same age my father came over. The mother, Clara, was a Nisei born in Stockton; her folks were also Kumamoto people. The Nakayamas' only daughter, Grace Sakioka, remembers the five housing barracks on the ranch. Each barrack had an individual kitchen, bedrooms, and then a Japanese bath that was shared with a neighboring family. The Nakayamas worked on our ranch for only a year; they then had to go to San Jose after the death of a relative.

My dad used to go just south of Portland, Oregon, to buy his garlic cloves for planting. And he used to have a big growing area on the Gilroy ranch just for garlic that would be used as seed. The first-generation Oregon seed was the most desirable. It was grown on fresh ground and cost at least 50 percent more than regular garlic. And then there was second-generation Oregon seed, which my father would raise in Gilroy and sell to growers who wanted good and dependable quality at a fair cost.

You couldn't fool a farmer by selling him something that wasn't at least second-year Oregon. First of all, you had your seed crop staked out on the ranch. Second of all, you could tell if it was first- or second-year by the size and quality of the garlic. Good garlic is dry and hard; you wouldn't want the early garlic, which has the skin falling off. Although the taste is not bad, this lower-quality garlic won't store as well.

Once the garlic was cured and topped, it was packed into field sacks and then stored. When we shipped the garlic back East in box cars, it would be resacked into new fifty-pound bags. Actually, since the garlic shrunk some as it dried during transport, we usually put more than fifty pounds, maybe fifty-two or fifty-three pounds, in each bag. Sometimes if the customers held them too long, the bags would weigh less than fifty, and they would insist on getting their money refunded.

Since we had to do all of this by hand back then, the smell stayed with us during the planting and harvesting season. Grace Sakioka remembers:

> *You get used to it. I remember when my father would go to the bank in town. They knew automatically that he worked in garlic because they could smell it.*

During the war, we left the garlic fields—for good. Some months before that, my father had started a partnership to sell garlic with Joseph Gubser Sr., whose father had come from Switzerland and gotten into dairy farming. I remember that I went with my father to San Francisco to a garlic distributor, the A. B. Hood Company, before the war to finalize the agreement.

According to *Smithsonian* magazine, Gubser Sr. was approached in 1943 by the Basic Vegetable Products Company about filling a large order from the government for dehydrated garlic. The company wanted Gubser Sr. to find twenty truckloads of garlic stored in warehouses by Japanese American farmers. Gubser Sr. refused to cooperate; his son, Joseph Gubser Jr., my high school classmate, instead located a farmer with a stockpile of garlic he was ready to dump. He bought the whole bunch for seventy-five cents per pound and sold it to the company at $2.75. This set him up in packing and selling garlic until his death in 1997.

New garlic growers, including Don Christopher, established themselves in the area in the 1950s. Now Gilroy is world famous for its garlic and its annual Garlic Festival, held every July.

Many have forgotten the early role of Japanese Americans in garlic, but some have not. In fact, there's a mural in town memorializing the achievements of pioneering garlic growers like Yoshio Nagareda and my dad. As it states in a 1997 publication by the Gilroy Historical Museum, "The largest [garlic] grower in the United States in 1940 was Kiyoshi Hirasaki of Gilroy."

CHAPTER

JAPAN PAVILION

About the biggest thing to come to San Francisco in 1939 was the World's Fair. In fact, the city actually built an island, Treasure Island, just for the event, also called the Golden Gate International Exposition. Like many fairs before it, the 1939 exposition featured Asian exhibits. What we wanted to see was the Japan Pavilion, whose construction involved a Northern California architect, George Gentoku Shimamoto, and two local San Jose men, the Nishiura brothers.

The Nishiura brothers, Shinzaburo and Gentaro, were well known throughout Santa Clara County. They had built most of the glass greenhouses in Mountain View, as well as a good number of houses for Japanese American families. The Nishiura brothers had constructed the San Jose Buddhist Chruch, as well as the Kuwabara Hospital in San Jose's J-town, the building where I often played outside during my year in the city. But they were more than carpenters or building contractors; they were artists, too.

That's why they were first called to build a Japan Pavilion for the Panama Pacific International Exposition in San Francisco in 1915. And then, close to twenty years later, they were brought on to work on the Japan Pavilion on Treasure Island. The pavilion, literally a large castle-like building with a three-level pagoda, housed exhibits on silk production, folk art, and trade. Inside was a room entirely made from mulberry trees and a *kozashiki*, a small traditional Japanese sitting room with a *tokonoma*, an alcove for flower arrangements and brush paintings.

Approximately 4.5 million visitors came to see the Japan Pavilion in 1939. The following year was the anniversary of the founding of Japan, and the Japanese government decided to reopen the pavilion during the last months of the Golden Gate International Exposition in 1940.

I was somewhat familiar with Japan because I had visited there once with my mother when I was a child. I remember stopping in Hawai'i and seeing the banana bunches from the edge of the boat. Once we arrived in Kumamoto, I began eating green pears, *nashi*, and I got awfully sick. The homes back then were primitive, with outhouses in the back. It's completely different today.

Even though my dad had left Japan so early, he was always interested in Japanese architecture. I would go with him on trips to Southern California to take a look at the Japanese tea house at the Huntington Library in San Marino, and later we would go to a club called Yamashiro, a replica of a Kyoto palace built in Hollywood in 1911. All along, he was looking at a style that he wanted to incorporate in a house for himself.

In 1941, with $23,000, he was able to realize his dream. About seven carpenters, led by the Nishiura brothers, came from San Jose every day for ten months to create a Japanese-style house next to our existing house. Actual pieces of the Japan Pavilion at the World's Fair, specifically the *kozashiki*, were brought in to create an authentic house. Wood panels carved by the Nishiura brothers were displayed throughout the living area. A Japanese-style *furo* was installed, and a local nurseryman, Kan Domoto of Hayward, designed a Japanese garden and pond filled with koi.

My father told the *San Jose Mercury Herald*, in a front-page story headlined "Japanese Farmer's Home Architectural Showplace" (29 September 1941):

> *I wanted a place to entertain my Japanese friends and also my American friends. A place that would show how things are in Japan.*

According to the article, Kiyoshi was first thinking about one room, then two, until he finally had a five-room house. I was barely aware of what was going on. I was graduating from high school that year and was more interested in how I could take off in my father's sedan. But I do remember the carpenters struggling with a fence surrounding our house and Japanese garden. In the style of Japanese traditional architecture, few nails were to be used in the construction in the house, so the carpenters had to create special detailed grooves at the top and bottom of the fence. It was hard work, and when it was finished, the fence separated our home from the acres of row crops that had enabled my father to build the house in the first place. Our Japanese house was finally completed a few months before December 1941.

CHAPTER

MR. RUSH

John Hardin Rush was an implement dealer in Gilroy and, more importantly, my father's best friend. He was about the same age as my dad—just a little bit taller and huskier. Both Mr. Rush and his brother Courtland had the same look: heavy-boned with brown hair. Mr. Rush, who was from Kentucky, sold farming equipment to growers like my father. His wife, Elsie, was a Rianda, the name of a local farming family. Mr. Rush and his wife had no children.

The ranch my father bought in 1932 was actually purchased in the name of an American-born relative. At that time, according to the anti-alien land restrictions aimed against Japanese and Chinese immigrants, you had to be a U.S. citizen to buy land. And there was no way an Issei could be a citizen before World War II unless they had fought in World War I. In the summer of 1941, my father switched over the land to a trusteeship in the name of myself and my sisters. But none of us were twenty-one years of age, so we needed a trustee to legalize the transfer. That trustee came in the form of Hardin and Elsie Rush.

Now I said that Mr. Rush was a good friend of my dad, but my dad also made a lot of money for him. In 1937 and '38, business was going well and my father had started to buy all kinds of equipment from Mr. Rush, and he'd push a lot of *Nihonjin*, or Japanese, into buying their machinery from him too. That's how it worked for a lot of the Japanese old-timers: if you made an introduction, you were putting your endorsement on that person. It's not like many *Hakujin*, who will shake hands with anyone.

In the late 1930s, my dad was one of the first to buy the latest diesel farm equipment and trucks from Mr. Rush. One year, all the growers on the ranch made a lot of money on garlic, so six of them went to San Francisco and bought cars from the wholesale warehouse. Aside from one grower who got a Packard—he wanted to be independent—everyone bought De Sotos and Plymouths. The farmers all drove back home together; it was funny to see the seven new cars parked in the yard.

Later, when Mr. Rush was establishing new dealerships in San Jose, my dad actually helped finance the construction of the new building. My dad was in between crops and didn't need the money at that time, so he loaned Mr. Rush the cash to put up the building, which was made up of sheet metal and stucco.

Clearly the connection between my dad and Mr. Rush ran deep. They weren't the kind of people who exchanged gifts and said a lot, but I know they were close. My dad had some other *Hakujin* friends, like Tony Silva—he was Portuguese and his family was in dairy farming—and Gordon Chapell—he came out to the ranch a lot to fix our water wells.

But things got rough after the bombing of Pearl Harbor. Although I never witnessed it myself, I know that people were calling Mr. Rush a "Jap lover." Mr. Rush told me that himself. And then there was the incident of the cherry blossom trees.

Right before the war, Mr. Rush had constructed a new building in Gilroy. As a token of friendship, my dad planted about five cherry blossom trees in front of his business. We heard later that someone had chopped some of the trees down during the war. We never found out who did it, and I guess we never will.

CHAPTER

HIGHER LEARNING

I didn't care to go to college. My mother pleaded with me to attend because education was important to her, but since I was raised on a farm, I thought that experience was more valuable than any kind of degree. I saw an engineering student return to town after graduating from a four-year college and he was soon back on the family farm himself. He just couldn't get another job.

As I was the oldest son, I thought that I should heed my mother's wishes, no matter what reservations I had. I entered the University of California at Davis, not far from Sacramento. I was practically the only one from Gilroy headed to UC Davis, but I wasn't the first Hirasaki to enter that university. My uncle Tokuzo, in fact, graduated from UC Davis around 1918.

My major was truck crops. You could also specialize in chicken raising and animal husbandry. I went through the first three or four months of school without studying. I thought I had it made, but then my classes began to get more textbook-oriented. With a little experience, you can get by on certain things, but eventually you have to study.

Truck crops are products that you handle in a large volume, like tomatoes, sugar beets, lettuce, cabbage, and green peppers. Crops that require special handling like strawberries and raspberries are row crops. Orchards—which produce crops like prunes, apricots, and walnuts—and vineyards are another thing altogether.

I lived in a private house with six other college students. Five were also Japanese American and one was Russian. As it turned out, we were more "Japanesey" in college than in high school. By that I mean that most Japanese hung out in groups with other Japanese. I was about the only one from Santa Clara County, but there were a lot from the south, including farming communities in the Imperial Valley. We would all go to Japantown in Sacramento and eat Japanese food. There was even a Japanese barber in Sacramento; she was a woman barber—a rarity—so there was a big curiosity factor there; everyone had to have their hair cut by her. Back in Gilroy, Filipinos had always cut my hair. I had never gone to a *Hakujin* barber. I never was a person who went against the wall to prove that I was just as good as everyone else. I never went for that.

I was receiving a pretty good allowance, so I was loaning five dollars here and five dollars there every now and then to friends. I was also conscious of how I dressed, so when I went to school I went with pretty new clothes. I had borrowed my dad's car for the semester, then a Buick Roadmaster, so I was also one of the few who had transportation besides a bicycle, although I did eventually have to return the car to my dad over Thanksgiving vacation.

I was back at the college house in December 1941. On Sunday morning, December 7, the Russian student woke me up. He was laughing and joking. "Hey, your countrymen are blowing up Pearl Harbor," he said.

All I could say was, "Where's Pearl Harbor?"

Right away I looked at the map. Later we were glued to the radio. I didn't know anything about politics. In fact, many of us Japanese Americans didn't. After the war started, however, many more became current-events people.

The faculty met with all the Japanese American students. They asked what they could do to help us. We really couldn't answer. We were farm boys and didn't make much noise; if you were too vocal, of course, somebody might knock you down. One student said that he was going to join the Air Corps and become a flyboy. He eventually did volunteer but was refused. Since I was the right age, I also thought about the draft.

My dad came to pick me up from school that December. I packed up all my belongings. I figured that I would be leaving Davis for good.

PART TWO: **WINTER**

Manabi Hirasaki, 522nd Field Artillery Battalion, Charlie Battery. **TOP:** *Nice, France, 1945.* **MIDDLE:** *Konigssee, Germany, 1945. Gift of George Oiye, Japanese American National Museum (95.158).* **BOTTOM:** *Camp Shelby, Mississippi, June 1943.*

TOP: *Manabi and Sumi Hirasaki standing in front of their 1947 Ford convertible. Mountain View, California, 1947.* **BOTTOM:** *Manabi and Sumi Hirasaki in front of his father's house (partially reconstructed from the 1940 World's Fair Japan Pavilion). Gilroy, California, 1947.* **RIGHT:** *Manabi Hirasaki and Sumi Iwata were married on January 19, 1947, at the San Jose Buddhist Church.*

Manabi and Sumi in front of their Gilroy house with son Mark Kiyoshi Hirasaki, 1950. Gift of Manabi and Sumi Hirasaki, Japanese American National Museum (99.85.1).

TOP: *Trucks at the Hirasaki Farms packing shed in Gilroy, California, 1950.*
BOTTOM: *Kiyoshi Hirasaki, founder, owner, and operator of Hirasaki Farms, at his desk, 1950.*

Our Farms brand crate label, used from 1947 to 1950.

Newspaper clipping featuring key staff of Hirasaki Farms, Inc., producer of Our Farms brand, late 1940s.

CHAPTER

ON THE MOVE

My father was picked up by the FBI on the second or the third round after the first group of Issei were taken into custody on December 7. The first ones to go were Japanese school teachers, Buddhist priests, and martial arts people. My father was high-profile at the time because, besides being the largest Japanese grower in the district, he was also the head of our town's Japanese school association that year.

I was there when the FBI came with the sheriff's deputies. One of the deputies was a good friend of my father. They acted like they were looking for something. After the men were finished, my father's friend came up to me and said, "If you got a gun, move it or get rid of it." Sure enough, we had a small shotgun in the house. The deputy, in fact, had seen it but didn't say anything. At that time, Japanese Americans weren't allowed to have shortwave radios or firearms. If we were found violating such laws, the authorities would have an excuse to arrest us.

Even though they didn't officially find anything, they still took my father away. We didn't know where he was for a few weeks; information was hard to get. If it had been the local police who'd taken him, we might have had a chance of getting some information on his whereabouts, but this was the FBI. Finally, two or three weeks later, he was able to write us a letter and tell us he was in Sharp Park, a military park ten miles south of San Francisco.

My mother was pretty coolheaded. Again, she had eight kids to take care of so she didn't have time to be upset. Luckily, our property was under a trusteeship with Mr. Rush, so we didn't have any monetary problems. On February 19, 1942, President Franklin D. Roosevelt signed Executive Order 9066, giving the military the power to arrest and detain anyone who was determined a threat to the government, even if there was no tangible evidence. This presidential order set up the foundation to create camps to incarcerate those of Japanese ancestry.

Many of our friends and neighbors moved east of Highway 99 to California towns like Reedley and Fresno. The government designated this area, in addition to the eastern portions of Oregon and Washington and a southern portion of Arizona, as Military Area Two. Officials said that this area would most likely be "free" and unrestricted, unlike Military Area One, which encompassed the western half of California, Washington, and Oregon, and other areas of southern Arizona. We in Military Zone One, on the other hand, would be placed into camps unless we could find new homes and sponsors in a free area in a matter of days or weeks. As it turns out, those in Military Zone Two were later also forced into camps, so some friends had made the move to Reedley and Fresno for nothing.

Unlike most Japanese Americans, we had a *Hakujin* trustee and advocate, Mr. Rush, who was instrumental in our time of need. "Let's go to Colorado because I got a friend there," he said. Actually it was a secondhand friend, a friend of a friend, but it didn't make much difference to us. We were open to Colorado because we had heard about its governor, Ralph Carr. He publicly welcomed Japanese Americans to come into the state. The governor of Utah, Herbert Maw, was also open. Most of the other governors in the area were negative or even hostile.

Mr. Rush and I first went over to Colorado in March 1942. We traveled by automobile. When we were getting gas in Nevada, a service attendant came up to Mr. Rush and asked him, "How come you're driving with an Indian?" Turns out that a lot of people in other parts of the country didn't even know what a Japanese American was, but that didn't stop them from disliking us.

We ended up in Grand Junction, located on the Western Slope, west of the Rockies, not too far from the Utah border. The Western Slope was known for its farmland, its peach orchards and fields of onions and tomatoes, much like the crops of Gilroy. But we weren't there to farm; it was just a place to move. "You've got to buy a house," Mr. Rush said, "or else you may get kicked around from place to place."

There were only a couple of houses for sale because of the war. One of the homes belonged to the assistant district attorney of the county, Mr. Grove, who had moved into a larger house in town. When we made a purchase offer, he was dumbfounded in light of the possible public backlash. "I'll get back to you in twenty-four hours," he finally told us. "I'll see if it's all right."

I guess higher authorities gave him the okay, because we were able to purchase the house. We felt pretty fortunate that a member of the district attorney's office was the former owner of our property; we figured he'd be out there helping us, too.

After purchasing the house, Mr. Rush and I returned to California. Then I practically turned right around to go back with two of my sisters, Fumiko and Michiko. The three of us were still teenagers, yet here we were traveling on our own. Mr. Rush advised us to leave the Buick Roadmaster in Gilroy—it was too fancy, he said, and sure to attract attention—so we traveled in my father's 1941 Plymouth business coupe.

We arrived in early March to get the house ready for the rest of the family. My mother and my other two brothers and three sisters got on a train on the last day Japanese Americans were permitted to leave California on their own. They almost didn't make it. By that time, Japanese Americans needed to get a permit with a valid relocation address to leave Military Area Number One, so Mr. Rush had taken my mother to San Jose to obtain a traveler's permit. When they arrived, a long line already stretched out from the permit office. At 5:00, the office was about to be closed before my mother's request could be processed. But as luck would have it, the manager, a Gilroy man named Jack Rocca, knew both Mr. Rush and our family. Rocca picked them off the line and personally issued my mother a permit.

So my mother, along with five children including my infant sister, took a train to Colorado. About midway, however, they missed a connection and had to stay overnight in a Japanese American boardinghouse in Ogden, Utah. A lot of other Japanese were stuck there too, including the Nakayama family, who had once grown garlic on our ranch. It was a good thing that this boardinghouse was there because how else could my mother have found lodging for the six of them in an unknown town? A motel might have been the most dangerous place for them.

Once they got to Grand Junction, we started to settle in. Our house had a large front room where I slept. It was a modern house with a dinette area instead of a formal dining room. Mr. Rush had been worried about the neighbors, so he had gone to see every household to tell them a Japanese American family was moving in. For the most part, they all were hospitable. An older German couple even brought over sauerkraut; that was the first time I had ever eaten the cabbage dish in my life.

But Mr. Rush did miss one person. He was a railroad man, specifically a train engineer, and a bachelor. When he found out about us, he raised a ruckus. But here's where my two brothers, seven-year-old Hisashi and six-year-old Shinobu, came in. His house was right next door, so my brothers would play in between the two homes. In time, he would call them in and give them candy. Pretty soon, they were always welcomed.

CHAPTER

GRAND JUNCTION

True to its name, Grand Junction was a large railroad junction with train tracks slicing right through one side. It was a rough-and-tumble town, full of bachelors who worked in the fields or with the railroad company, the Denver and Rio Grande, or DRG. Most of the railroad men wore coveralls soiled with soot.

There were always some longtime Japanese Coloradans there: the Furukawas, Hayashis, Hases, and Odas. The Furukawa family grew onions. Mr. Hayashi had a farm, and his son operated a restaurant that served only American food. Most of their customers were railroad men.

My sister Michiko remembers that an older man in a Grand Junction grocery store once got in the way of our mother and wouldn't move. Mom looked right in his face, smiled, and said, "Hello." It was then that he moved. Mom was smart in that way; she wasn't one to look for trouble.

A lot of the Japanese old-timers in Colorado didn't know how to take us newcomers. The *Rocky Nippon*, a Japanese American newspaper published out of Denver, carried a series of articles and letters to the editor about us Californian "evacuees" in 1942. In March one writer considered the impact of large numbers of Japanese Americans moving into areas where others had been living "peacefully for many years":

> ...let us imagine a thousand, two thousand, or even five thousand or more Japanese making their homes in your locality alone; what reaction would you and other present residents have?
> (Rocky Nippon, 3 March 1942)

As it turned out, less than two thousand Japanese Americans moved out to the entire state of Colorado during this window of time. In fact, only 4,889 "voluntarily" relocated from Military Area Number One to other parts of the country by the deadline of March 29. Later, after Japanese Americans were forcibly removed into camps, thousands more were allowed to relocate into "free zones" like Colorado if they passed a security clearance and qualified for either college or outside jobs.

One Californian who called himself Sugah-Foot, referring to the sugar beet farming in Colorado, complained about the weather and the girls in his new home in a letter published in the *Rocky Nippon*. The Nisei Coloradans didn't take too kindly to his comments. One reader wrote:

> *In ending I want to say that we Coloradoans [sic] do like the evacuees, but wish you would get things straight. After all, we didn't get down on our knees and beg you to come here. Some of you came despite the fact we asked you not to. You coming in and in some cases your behavior has made it hard for many of us already living here. We are the ones who pay for all your thoughtlessness. And in face of all of this, when you hear evacuees complain and read articles like Sugar-Foot's [sic], it does make me very, very mad!*
> (Rocky Nippon, 3 August 1942)

I, on the other hand, didn't make too much noise. I wasn't aggressive; rather, I stayed quiet. With my father gone, I had to work at home, doing the outside chores like fixing the septic tank when it went out and taking care of the garden. I pretty much stayed at home, except when I went uptown for groceries.

During that first year, I did day labor in the peach orchards and in an onion repacking warehouse. In the summer we filled harvesting sacks and in the wintertime we repacked the onions into new bags to ship. I also worked part-time in the summer at a bunker, icing refrigerator cars at the DRG railway. I was given a large pitchfork with a sharp tongue and told to get on top of the railcar. Standing about eight to ten feet above the ground, I waited for a three-hundred-pound block of ice to come down a chute from the ice storage warehouse. As it passed by on a ramp toward the top of the railcar, I had to split the block of ice in half with the ice fork. The pieces could then fall right into the bunker. It was dangerous work and not too much fun.

I also helped to unload the railcars for a wholesale hardware company. Because of this work, I was able to apply for my Social Security card in Grand Junction. My mother, on the other hand, kept egging me on to go to school. So in the fall I enrolled in Mesa Junior College. It was a small school; there were maybe half a dozen Japanese American students. We hung out in a mixed group and sometimes went over to the local bowling alley and the coffee shop next to it. I didn't do any serious studies and I felt a sense of laziness start to seep into me. I didn't want that feeling of laziness to go any further; in that way, I was like my father. I didn't want to sit around—that was the worst thing to do. I needed to make a change, and in time, I did.

CHAPTER

BISMARCK, NORTH DAKOTA

In time we learned that my father was being held in a Department of Justice alien detention camp just outside of Bismarck, North Dakota. At age forty-two, my father was still on the younger side. Since he had come over to the United States at an early age and was fluent in English, he was used many times as an interpreter, leader, and group contact.

The detention camp was located in Fort Lincoln, south of Bismarck. Italian and German seamen who were stranded in the United States were first incarcerated there in the spring of 1941. Later, the Japanese like my father came in. There were other Justice Department alien camps in Missoula, Montana; Santa Fe, New Mexico; and Crystal City, Texas.

My father was a bear for work. He didn't like to lay around, so when the call came for sugar beet thinners in Montana, my father volunteered and was ready to go. Sugar beet thinning was basically stoop labor. All you needed was a strong back and a short-handled hoe. Sugar beets were grown in rows from seeds during the spring. They grow thick, so workers had to chop in between the plants, leaving six to eight inches separating each row. Then, when the sugar beets were ready in fall, they would pull each one, cut the top off, and stack them in a row.

The War Relocation Authority, the federal agency in charge of the removal of Japanese Americans from the West Coast, wanted my father to organize a crew to work on sugar beet farms in the Montana area. My father was sent there first to check out different farms and make arrangements for the workers' housing and pay. Farmers told my father that they would do everything possible to make it comfortable for the Issei workers, and my father took them at their word.

But once they began work, the farmers didn't keep their promises. My father felt pretty bad because he had recruited so many workers and they weren't being taking care of. I'm not exactly sure how bad the conditions were, but they must have been terrible enough for my father, who never spoke ill of anything, to mention the little that he did. He also didn't say much about Bismarck other than that he didn't like how the authorities counted them as if they were sheep. The prisoners were just numbers, he told me. That was about all.

He was one of the first Issei in Bismarck to receive a hearing for his release. Here again, Mr. Rush came out and testified on his behalf, and as a result, my father was freed in September 1942. He was one of the first Issei to get out of that detention camp.

I was the one who picked him up from the Grand Junction train station. The station was simple, just a platform with a row of benches. When my father got off the train, he was smiling. Once at home, although he was very quiet at first—then a little short-tempered and irritable when he wasn't doing anything—Bismarck really hadn't changed him much. He was still the same man, taking everything in stride.

CHAPTER

ENLISTING

My father didn't waste any time when he came to Grand Junction from Bismarck. We did some tomato and sugar beet day labor work in the area before he leased some land to grow onion seed. He couldn't produce garlic because Grand Junction was too cold in the wintertime and not hot enough in the summertime, plus it would have been too difficult to obtain garlic cloves for seed in the first place.

One day he went into town to go shopping. He was carrying an armload of groceries when he saw someone from our hometown—Tony Silva, a Portuguese man who used to drive a tractor and operate other equipment on our Gilroy ranch. My father was so excited that he dropped all his groceries and hugged Tony. Tony's family had owned a dairy farm next to our old ranch. During the war, Mr. Rush had leased part of our farm to Tony, who was trying his hand in vegetable growing. He had come to Grand Junction to visit us.

In time my parents were communicating with other folks from Gilroy who were in camps, mostly in Poston, Arizona. The Nagaredas were in Poston I. The family stayed in close communication to the Pieters-Wheeler Seed Company; my friend Hiromi even went to check out its operation in Caldwell, Idaho. But the acreage was too small to support a large family, so Hiromi eventually went to Chicago to become an auto mechanic. One of his roommates was Noby Yamakoshi, who ended up starting one of the largest catalog design and production firms in the nation. My friend Moose Kunimura had also been in Poston I, where he was captain of the Poston High School football team. Later he played football for the U.S. Army during the war and missed being placed on the front lines.

I was also thinking about military service. At Mesa JC, I saw all the young men waiting to be drafted. My junior college also had preflight training for naval pilots. The only ones staying behind were those who were 4F, physically unfit for service. No young man wanted to be 4F, and I was no exception. In January 1943 I decided that I was going to volunteer to the draft board and join the Army.

My parents didn't say much. My mother wanted me to stay in college, but I went to the draft board in Grand Junction on my own anyhow. I filled out a form and later was told to go over to Denver for my physical. Quite a few of us took the bus from Grand Junction to Denver. Since we were all entering the Army, there was a lot a camaraderie on that trip. Nobody seemed to notice that I was Japanese. In Colorado, Japanese Americans were often thought to be American Indians. In fact, some people I know were not served alcoholic drinks because Colorado law at the time barred American Indians from buying liquor.

After I'd completed my physical, a recruiter from the Marines came up to me and asked, "Do you want to join the Marines? With your physical, you are able."

"Go ahead," I said.

Then he left to consult with someone and returned, saying, "Sorry, we can't take you."

A Navy recruiter then approached me the same way. I didn't want to join the Navy because I knew a lot of enlisted men became waiters and cooks. But I told him to go ahead and try it. And, as was the case with the Marines, I was denied because I was Japanese. At that time, I didn't know that Japanese Americans were considered 4C, or enemy aliens, and I guess the Denver recruiters weren't aware of it either. I was told that I would have to wait.

Another Grand Junction boy named Hase, however, made it in because they didn't realize that his last name was Japanese. His mother was Mexican. The Marines sent him clear down to San Diego before the officers realized that he was part Japanese. As a result, he was sent back to Grand Junction.

Around February and March we began hearing about the 442nd Regimental Combat Team. It was a segregated unit made up of all Japanese Americans. I liked the idea of a segregated unit because it would feel more like home. I felt as though I could depend on my friends more.

Since all my papers were processed, I was one of the first mainland boys to be called into service. In April 1943 I was sent to Camp Shelby near Hattiesburg, Mississippi, for basic training. Colonel C. W. Pence of the 442nd Infantry then sent a letter to my father dated 23 April 1943:

> *Dear Mr. Hirasaki:*
>
> *You have given a soldier to the Army of the United States. Private Manabi Hirasaki has arrived here safely, and I am happy to have him in my command.*
>
> *To you I extend my congratulations. By your sacrifice you have enabled him to enlist voluntarily, and become a symbol of the loyalty and patriotism of our Japanese-American population. Without compulsion or persuasion, he made the brave and manly choice to fight for the American way of life. He freely chose to exercise the responsibilities of his citizenship.*
>
> *President Roosevelt has stated: "Americanism is not, and never was, a matter of race or ancestry. A good American is one who is loyal to this country and to our creed of liberty and democracy."*
>
> *I am sure that you are proud of the soldier you have given to us. With him, and others like him, we shall make a glorious record for the Japanese-Americans in our country.*

CHAPTER

Camp Shelby and the 522nd Field Artillery Battalion

When I got to Shelby, I told them my first choice was to be assigned to the 232nd Combat Engineer Company of the 442nd Regimental Combat Team. I thought that would be best for me based on my experiences with machinery and heavy equipment. But the Hawai'i boys, many of whom were already in construction, pretty much had the engineering group filled. It was the only unit in which all the officers and enlisted men were Japanese Americans. Most were from Hawai'i, but a few were from the mainland, like me.

My second choice was artillery and my third, infantry. At that time, I didn't know how dangerous the infantry was; maybe I wouldn't have come home if I had become an infantryman. There were six of us from Colorado. The officers lined us up against the wall. The three tallest boys were to go into the artillery; the shorter ones, the infantry. At five feet eight, I was the tallest one.

I first went to the field infantry headquarters at Camp Shelby, and three days later, I was in the artillery battalion. The 522nd Field Artillery Battalion was the support unit of the 442nd Regimental Combat Team; we would provide the firepower. Almost 650 men were part of the 522nd, and within the 522nd were smaller groups: Headquarters Battery (160 men), Service Battery (86 men), Medical Detachment (16 men), and then the three firing batteries, A, B, and C (125 men each). These last three were also known as Able, Baker, and Charlie, respectively; those were the words soldiers used over radio communications during World War II for the first three letters of the alphabet. I was assigned to Charlie Battery.

Within Charlie Battery were approximately ten different sections, including the gun and supply sections. Named to the wire section, I was responsible for helping to aid communications between headquarters and the guns.

Since I was the first rookie from the mainland in a group of soldiers from Hawai'i, they didn't know what to do with me. They assigned me to a hutment, or a wooden barrack, where non-coms, or non-commissioned soldiers like sergeants and corporals, were staying. The Hawaiians would look at me and wonder why a private was getting special treatment: "Who does this *kotonk* (mainland boy) think he is?" In the middle of the night, the non-coms would get up and go to the kitchen for a snack—something other enlisted soldiers could never do. I'd go right along with them. Unfortunately, that only lasted about a week or so before I was moved to another hutment.

As I said, in April 1943 the 442nd was made up of mostly Japanese Americans from Hawai'i. I noticed that the Hawaiians would come out and say what was on their minds; verbal and outgoing, they wouldn't hold back. The Nisei from the mainland, on the other hand, wouldn't go so far. The difference was that we were the minority in the states and they were the majority in Hawai'i, I figured.

In the beginning there was a lot of tension between the Hawaiians, or Buddaheads, and the *kotonks*. We didn't know how to joke around; everything was taken as fighting words. For instance, one Seattle boy chewed his gum with a snap. The Hawaiians didn't like how he chewed his gum and that led to a fight.

I nearly got into a fight myself. It was a hot, muggy day at Camp Shelby, and a Hawaiian soldier came up to me and told me to scratch his back. He had an itch in a spot that was hard to get.

I didn't like his attitude and the way he approached me. "Go get someone else to scratch your back," I said. The guy wanted to fight me after that. He waited for me in front of my hutment, but I ignored him. There was no use in fighting over something like that, I thought. Later I found out that the soldier—his name was Mel Sakata—was a boxer. He even boxed professionally after the war, so it was a good thing I didn't tangle with him.

A lot of the Hawaiian boys were from plantations, while some of the mainland guys were from families like mine that operated their own farms. Many of the Hawaiians didn't get a chance to pursue an education and they didn't like how the mainland boys spoke. They spoke pidgin English: "I stay come, you stay go." Pretty soon pidgin English was getting into all of us.

There was a lot of competitiveness among us. In terms of the foot races and exercises, we wanted to be just as good as or better than the next guy. We were all young kids and we had a lot of get-up-and-go. If you slacked off, everyone would tease you. In this sense, the competitiveness was good because we knew we could depend on each other.

I myself never went to visit the Japanese American concentration camps, but some of the Hawaiians did. Members of the 442nd, in fact, went to the Rohwer and Jerome concentration camps in Arkansas, which was fairly close to Hattiesburg, Mississippi, and Camp Shelby. The Hawaiians were elated to take a trip and on the way there played Hawaiian songs on ukuleles and sang. But when they arrived at the camp and saw the barracks and the conditions there, the soldiers were shocked. On the way back to Mississippi, no one was singing.

I went back to Grand Junction on furlough. By that time, my father had brought a lot of Gilroy people, including Bill Yamano, Shig Yamane, and the Kishimura family, out of camp into the Western Slope of Colorado. The truck crops grown around Grand Junction were similar to the ones we dealt with in Gilroy. To get out of camp, Japanese Americans needed an address to go to, and my father was able to provide that. Usually the head of the household took a week to look things over so families wouldn't come to Grand Junction blind. During the war, some people lost their property and farms back in California because they didn't have everything paid for.

So it was that Colorado, for a while, would be our new home. In the Army, in fact, my nickname was Colorado because that was where I came from when I enlisted. It didn't matter to anyone that I had actually spent most of my life in a California town called Gilroy.

CHAPTER

MANEUVERS

In the fall of 1943, the 522nd went to Louisiana to conduct some maneuvering exercises and we stayed there through the holidays between Christmas and New Year's. This was the first time that our artillery unit was working without the infantry of the 442nd Regimental Combat Team. The weather was so bad and the maneuvering ground in Louisiana was so old that there were holes everywhere about waist deep. You had to watch where you stepped.

As low-level soldiers, we didn't have the problems faced by the artillery officers. They had a lot of equipment to move, and they had to make sure the timing was just right. We only had to follow along in the maneuvers. We didn't understand what was going on at all times, but then we weren't supposed to.

The wire and radio section had a switchboard and from there we laid wire out to headquarters. There were about nine of us in the wire section and three in the radio section. For the most part, the Army didn't like to use radio because the enemy could easily intercept messages. Telephone back then was a better mode of communication for close work.

Connected to our section was someone who grew up near Gilroy, Shiro "Juggie" Takeshita. Juggie was from the town of Salinas. His father, Masamoto, an immigrant from Kagoshima, had been a tailor who owned a shop in the middle of town. Juggie had four brothers and one sister; they were all sent to camp in Arizona. His oldest brother became an optometrist and stayed at the Poston II camp to take care of the internees. Juggie and his three other brothers all went into the Army.

Juggie joined Charlie Battery as a machine gunner on our wire truck. We all had to learn how to shoot a machine gun, but men like Juggie had more extensive training on it. He had to learn how to take apart and put together a gun while blindfolded. When the machine gun wasn't on the wire truck, he would be manning the gun alongside our four howitzers. A howitzer, also known as a 105mm, was used to shoot projectiles. Juggie explained:

> We needed two machine guns in the battery for every time the howitzers were set up in place. In order to protect the flanks, they had one machine gun on one side and another machine on the other. Usually I'd be on the right hand side. [Later] we'd take the machine gun off the mount and set it up with the tripod in the dugout.

During our maneuvers, each soldier had to team up with another. Everyone carried one half of a light-canvas pup tent. You and the other guy would then pitch and share the tent. I usually ended up with a Hawaiian named Thomas "Tom" Takano. We were both in the wire and radio section. He was quiet like me and kept himself neat; he was never sloppy. After the war, he went to the University of Hawai'i on the GI Bill, took ROTC, and went back into the service. He eventually became a lieutenant colonel.

In March 1944 we were told that General George C. Marshall was coming to Camp Shelby to conduct an inspection. Our regiment prepared to participate in a big parade for the general. Rumors also went around that we would be going overseas soon. More than three thousand participated in the inspection. We all looked sharp, and I was proud to be a Japanese American. A few months later, our orders came in. The 442nd Regimental Combat Team, including the 522nd Field Artillery Battalion, was to leave for Europe in May of that same year, 1944.

CHAPTER

NAPLES

The night before we left for Italy, we stayed at Camp Patrick Henry in Virginia. Juggie was there and remembered how the USO (United Service Organization) had a dance for us:

> *Knowing how the Hawaiians are—they don't take any guff. One of the fellows says, "I want to dance with that girl," and cuts in on a [Hakujin] guy...Well, the* Hakujin *guy didn't like that, and first thing you know there was a riot. And pretty soon that whole USO was a shambles. And they finally had to call out the officers of the day, and they came with weapon carriers and whatever to try to stop it. But anyway, they finally got everybody back to their barracks. And to us, we thought that was that.*

The next morning, the 442nd, 522nd, and all the officers went out on one boat from Newport News, leaving about seventy-five of us Charlie men behind. They didn't have room for us, so we were assigned to another ship with mostly *Hakujin* Air Corps soldiers. As we got into the boat, we noticed a lot of fellows with patches on their hair, little cuts and bruises from the fight the night before. "Uh-oh, we're really in trouble because we're outnumbered this time," we thought. But there were no repercussions.

It took us about twenty-six days to cross the Atlantic because instead of going directly to Naples, we went around Italy through the Adriatic Sea to the town of Bari, on the heel of boot-shaped Italy. At night we stayed on deck to look out for German U-boats, and when we got a chance during the day, we slept downstairs on bunk beds. We arrived in Bari on May 28, 1944. From there, we took a train ride on small boxcars called "40 and 8s" because they could each hold forty men and eight horses. We finally got off the train at Bagnoli, just outside of Naples, and met the rest of the 442nd and 522nd.

It was the tail end of May—orange and cherry season. I remember eating red oranges for the first time. Their red meat and red juice was so good. I didn't go into town much; I was getting ready to go up north in a few days.

On June 6 we went to Port Nisidra to travel a short distance to Anzio Harbor. We had to climb twenty-foot nets to get into the ships. While I was climbing, I heard someone holler, "Anybody from Gilroy?" Gilroy? I thought that maybe they were saying Kilroy, that popular cartoon of the time. But when I got to the top, I met up with Ronald Hadley, the son of the principal at Gilroy High School. He was an ensign and had run into another local friend, Gassy Sakaguchi, who had assisted the Nishiura brothers in moving parts of the Japan Pavilion from Treasure Island to my father's Gilroy ranch before the war. Both Hadley and Sakaguchi were waiting for me. That evening I got to eat in the officers' mess hall, but a day later we arrived in Anzio Harbor, from which the hometown friends went their separate ways through Italy.

CHAPTER

MINEFIELDS

AND FORWARD OBSERVERS

Frankly, when we first moved into the battlefields of Europe, I didn't know whether to be scared or not. We didn't have the same experiences as the 100th Infantry Battalion, which had served on the front lines of Italy since August 1943. The 100th, made up of Japanese American men from Hawaiʻi, had been together since the summer of 1942, when they arrived at Camp McCoy, Wisconsin.

The most dangerous job within the 522nd was to be in a forward observer party. The forward observer unit was usually comprised of an officer scout, a radio man, and a wire man, who served as assistant to the radio man. The forward observers located and directed fire on targets. We were positioned with the leading infantry patrol, so if we didn't read our maps correctly—the most important job—shots could fall short and our own troops might be hit by friendly fire. There were times in which I was with the forward observer unit on a hill, wondering if shells would make it over my head.

Headquarters would give us orders regarding fire direction. Usually they would call for just one gun to show the round of smoke; from the smoke, they could judge where to adjust their target. The number of powder bags placed in the gun determined how far the shells would travel, so you wanted to make sure that the man loading the bags could count. There were also various types of ammunition: armor-piercing shells for tanks, personnel shells that burst in the air, smoke shells, and color-coded shells. The Germans had what we called "screaming meemies," which emitted a shrill noise from the gun. Sometimes the sound would hurt you more than the shells.

Since it was so risky, we would take turns as forward observers. One night a group of us on a forward observer mission received orders to move deeper into Italy. It was then that I walked through my first minefield. Usually the groups before us would mark the safe path, but there was always the chance you could take a fatal misstep. Some people marked the trail with toilet paper—we all carried toilet paper, after all. Some lucky outfits had minesweepers that could detect metal underneath the ground, but we had to depend on only our five senses.

When daylight hit us, we felt the tiredness soak into our bodies. Some of us rested against an abandoned stone barn, our faces toward the sun. After some time, I walked away from the group only to hear an explosion behind me. Sure enough, a German mortar shell had hit the barn wall. Some of the infantrymen were hurt but no one was killed, fortunately. We learned an important lesson: next time we should spread out and not be all together as a sitting target.

The wounded got patched up and the rest of us moved into new ground. After walking many miles, we took a break. I was resting on the side of a hill when I noticed that a jeep was driving up to a barn in the distance. At that time, I didn't know what a German jeep looked like, but I thought that this jeep had a funny shape to it. I alerted the fire direction officer, who fired into the barn. Sure enough, German soldiers jumped into the jeep and took off.

At night we heard a lot of movement in the valley. It didn't come from Americans, as the talking and hollering was in German. We fired quite a few shells into the area and, like the German soldiers in the barn earlier, they left in a hurry.

Those experiences initiated me into life in a war zone. We didn't have time to be frightened, and many times when we were, it was too late. I had to focus on my main responsibility: to lay down wires for communications. Back at training camp in Mississippi, we trained with heavy telephone wire covered in black material, but in Europe, we sometimes worked with German wires that were encased in plastic. Lighter and color-coded either green or yellow, this double-stranded wire was a lot easier to use. We would look for these telephone wire spools in abandoned warehouses and sometimes hooked into preexisting lines. The spools were huge; we could only transport about three spools in one jeep. I'd usually run alongside the jeep, laying down the wire, while another man in the jeep would unwind the wire from the spool.

There were times in which we required extreme firepower. We were providing support to the 442nd Infantry and other attached units of the 34th Infantry Division along the Italian coastline when we reached Hill 140 in July 1944. More than forty-five hundred rounds were fired during a twenty-four-hour period in our effort to drive the Germans out of the area. The supply crew couldn't bring up the shells fast enough. It was a real battle.

I'll never forget when we were in Bruyeres, France, providing direct support for the 442nd, which was attempting to rescue the 141st Infantry Regiment, also known as the "Lost Battalion," composed mostly of boys from Texas. It was about this time that I saw many dead Germans along the roadside. Their corpses were frozen solid and American soldiers stacked them on trucks and trailers like pieces of cordwood. I was laying down wire in the road from headquarters in Bruyeres back to C Battery. When you are working under these conditions, you want to make sure that the wire is well secured to something stationery, like a telephone pole. On the other hand, you don't want to stay in the road too long because the road is an open target. This road curved at one point, so I tied a wire around a post and continued on with my wire work. Then I heard a loud explosion. A wire man from another unit was blown through the air. He had been laying wire in the same bend of the road and had tripped a land mine. I had worked the same spot just minutes earlier.

We in the 522nd were lucky. Only two men in our battalion were killed—one accidentally and the other killed-in-action, or KIA. The KIA was Nobuaki Tomita from Baker Battery, who suffered fatal injuries from an artillery shell on November 6, 1944. At the time, he had been a forward observer.

CHAPTER

MP in Monte Carlo

The best meal I remember we had was Thanksgiving dinner in France, near the French-Italian border in November 1944. We ate on a bluff overlooking the city of Nice and wondered where we were going next.

The 442nd was pretty mangled and the soldiers were ordered to R&R, or rest and recreation, in Nice while waiting for replacements. Then came a call from the first sergeant's office: two men from each company were needed to serve in the military police, or MP, company. They were taking volunteers, and George Suenaga and I went from C Battery.

I expected that I'd be guarding prisoners of war, but as it turned out, I ended up guarding our own GIs who got into trouble. These were members of the 442nd and surrounding units who drank too much and fought with one another. It was a thankless job.

There were some perks, however, that came with MP duty. First of all, we took care of the post exchange, PX, which sold goodies like candy and cigarettes. Every soldier was allotted a certain number of supplies if they had the money. Many, however, were broke, so I sometimes loaned friends money for these extras. If they couldn't afford it at all, I bought their allotment for myself.

As a member of the military police, we were also able to patrol on jeeps through areas off-limits to other GIs. Monte Carlo, located in Monaco just northeast of Nice, was one of these places. In Monte Carlo we were able to purchase soda water, Coke, 7-Up, and bakery goods. We even saw all the shows, including the French hit *Gigi*, and went to a lot of dances. We didn't miss a thing.

But this time of fun was short-lived. On my twenty-second birthday, March 8, 1945, we were on our way out of France and heading into Germany.

CHAPTER

DACHAU

When we separated from the 442nd to go into Germany, we of the 522nd felt like orphans. We were now attached to different units that we weren't that familiar with. But, whether we felt comfortable or not, we continued to do our jobs.

Once we hopped over the Rhine River, we moved fast—faster than even the infantry units since we traveled in trucks and jeeps. Sometimes we moved two, three times a day and all throughout the night. Every time the infantry moved, we would have to leapfrog up. Our shooting range was now much shorter: about five miles. The Germans were retreating and we were breaking loose.

Just before the war ended in Europe in May 1945, we moved to a thirty- to forty-mile area called Dachau, northwest of Munich. I was laying down telephone wire with another group, so I was dealing mostly with battery and battalion headquarters. But other men who were part of some forward observer teams traveled through a wider area. They were the ones who came across the Dachau death camp.

Later we all heard about the ovens and gas rooms in which Hitler had killed Jews. It was so hard to believe. Being a hometown boy from Gilroy, I didn't even know what a Jew was. To me a *Hakujin* was a *Hakujin*; I didn't know the difference between a Jew and an Anglo-Saxon. In fact, when I first saw a Jewish Holocaust survivor, I thought he was a German prisoner of war.

We weren't supposed to give the Jewish survivors any food because they were weak and their stomachs could not handle anything solid. One day, after cleaning our mess kits, I noticed some Jews were picking leftover food out of the soapy dishwater we had thrown out. They were that hungry.

After Germany surrendered on May 7, 1945, five hundred and fifty of us in the 522nd were spread all over the country. Many times we set up our communication switchboards in personal homes. Once, we took over a German chicken ranch and for the next few days cooked chicken as many ways as you can prepare it— barbecued, stewed, roasted. All we ate for a few days was chicken and eggs. Chickens were so valuable on the war front that we even saw other 442nd men carrying around live hens or tying them up like pets. Fresh food and vegetables, as well as canned meat like Spam and corned beef, were valued at a premium. One time I even saw a soldier pick some sugar beet leaves, thinking they were greens. He must have been surprised when they turned pitch black when stir-fried over a flame.

On our way from R&R, we stopped by the main concentration camp in Dachau. It was about August or September, so everything was fairly cleaned up. The only thing left were some leather shoes laying on the ground. We saw the gas room and furnaces, and I took a roll of pictures. I even laid down in one of the caskets and pretended I was dead. Later, when I saw the developed pictures, I wondered what I could have been thinking and threw the film away.

CHAPTER

Coming Home

By June 1945 we had moved into a German town called Donauworth to begin our occupation duties. Some went to school and others participated in sports and other recreational activities. We all were waiting to be either reassigned to another unit or discharged. We could only be discharged if we had earned enough points through a system in which officers and soldiers who had been in the service longer received more points. The Hawaiians had the most seniority and the most points because their months of basic training on the mainland counted as overseas service.

Mainland rookies like me, on the other hand, were assigned to different units. In September 1945 I was among one hundred and thirteen mainland boys from the 522nd to be transferred to the 53rd QM Base Depot near Nuremberg. Three of us had been the first ones to join the Hawaiians in Camp Shelby back in April 1943. I had been together with my comrades for more than three years, and we were as close as college classmates. These people knew me better than my own folks knew me.

I figured that I was doing my share for this country. We weren't thinking about ourselves but the generations to come. I was lucky because I didn't have to sacrifice my life. Combined, members of the 100th Infantry Battalion and 442nd Regimental Combat Team received more than eighteen thousand individual decorations, including twenty-one Medals of Honor, becoming the most decorated U.S. military unit of its size and length of service. But the costs were deep; casualties numbered 9,846. I had no idea when I enlisted that all of us might not be coming back.

Finally, in January 1946 I received my discharge orders with thirty-five to forty other Nisei. It was long after the war, so there was no homecoming for us. I spent two weeks in Chicago with my buddies before heading back to Gilroy, where my family had returned. My dad wasn't the kind of person to show a lot of emotion, so when I returned to the ranch, his main message was "Let's get to work."

CHAPTER

Hirasaki Farms, Inc.

Gilroy and the outlying areas had changed much when I returned. Strangers had come into town, while old friends had joined us on the ranch. Due to a Supreme Court decision rendered in December 1944 and a reversal in government policy, Japanese Americans could have returned to California as early as January 1945, but many hesitated. Hiromi Nagareda, in fact, sent a letter to the seed company that was operating his family ranch. The original founder, Lin Walker Wheeler, and his wife had died in a tragic airplane accident during the war, so it was the new president of the seed company who wrote back in a correspondence dated 19 January 1945:

> *Miss Morita wrote to me recently saying that her father would like to come back to California now that all restrictions are removed. She asked me the opinion which prevailed in this locality regarding the change in regulations. Before answering I checked up as far as I could and found that most people are quite bitter and I would advise anyone particularly an National [sic] or women to remain away from here for the present. The Philippinos [sic] are most bitter and of course the average person can not forget*

Pearl Harbor. It would seem to me that since there is great shortage of houses your people would be better off for the present to remain where they are.

Hiromi eventually left Chicago and picked up his folks in Poston. They first came back to San Jose, where their uncle had resettled, but there was no place for the rest of the family to stay so they came to my father's ranch in Gilroy at the end of 1945. The Nagaredas were a large family, numbering more than a dozen people. Their new temporary home was our seed mill, which had been a cheese plant prior to the 1930s. It had ample room and thick concrete floors, and Hiromi and his brothers had constructed partitions to create different rooms.

Other farmers in the neighboring area also took in Japanese Americans returning to the West Coast. Many came to Pajaro Valley, an area that stretches from Gilroy in the east to Monterey Bay in the west. The Uyematsus, a family who had farmed strawberries and tomatoes in Long Beach, California, before moving to Chicago during the war, came to Watsonville, the largest city in Pajaro Valley on the recommendation of a family friend. They lived in a camp on the ranch of Thomas Porter, a pioneering strawberry grower. A total of a dozen families resided in the barracks, which were divided into different rooms. The Uyematsus had their own kitchen and sink.

As subcontractors at the Porter Berry Ranch, the Uyematsus learned how to grow strawberries, Watsonville-style. At the start, they were responsible for about three acres. And then it grew until the Uyematsus became full-fledged strawberry growers on their own. Sho Kobara, whose father had to abandon ripe strawberry fields in Salinas in 1941, also came to Watsonville after serving with the U.S. Occupational Forces. He was able to get a job on an apple orchard operated by Gerald Sheehy, who also farmed strawberries.

The other big farm outfit was the Driscoll ranch. The Driscolls, who had been growing strawberries with their relatives, the Reiters, since the mid-1800s, had depended largely on Japanese American subcontractors before the war. The Driscolls had provided housing, land, fertilizers, and supplies, while the contract growers took care of all the labor. During the war, the Driscolls were one of the few growers to have anything to do with strawberries. In fact, according to Kazuko Nakane's research on the Pajaro Valley *(Nothing Left in My Hands: An Early Japanese American Community in California's Pajaro Valley)*, there were only 8 acres of strawberries farmed in the area in 1945, compared to 422 acres in 1940. The number of farms dropped from seventy-one to ten during the same time period. Most of Driscoll's wartime strawberries were farmed in Salinas, where the fruit went straight to a processing plant to be frozen or manufactured into preserves. Once the war seemed to be coming to an end, Japanese Americans from the Poston concentration camps were recruited to work on one of three Driscoll ranches. About twenty-five families came out and lived in barracks that had been moved to Santa Clara County. The Driscolls must have had at least one hundred cabins for families stretching from Morgan Hill down to Santa Maria.

During the war, our Gilroy ranch had been leased to a farmer named Henry Allamand. Mr. Rush and his wife, Elsie, kept meticulous books and had even paid off the mortgage, so we were in good financial shape, much better than most Japanese Americans. When the lease officially ended, my father was ready to farm. He started off with cucumbers, a spring crop. They grew fast, stretching out like fingers from vines. By early summer, we could start harvesting. We made more money on small cucumbers, which were used to make fancier pickles. I drove the cucumbers into the cannery two times a week, and then before we knew it, the cucumber season was all over.

In the fall we began planting lettuce and then tomatoes, and finally celery and fall peas. In 1948, with myself and three sisters all close to twenty-one years of age or older, we had officially incorporated as Hirasaki Farms, Inc. On paper I was the president and my sister Mineko was vice president and treasurer, but we all knew that my father, officially described as the general manager, was the one in charge.

CHAPTER

13

ROMANCE AND PICKLES

In January 1946, shortly after I returned to Gilroy, I went to a Japanese American Citizens League (JACL) dinner at the Lucas Restaurant in Santa Clara. It was my first postwar Japanese American get-together in the area. That's where I met a Mountain View girl named Sumi Iwata.

I had actually dated Sumi's older sister Maki once before the war. Their father, Daemon, originally from Wakayama, Japan, had been a flower grower. Mountain View was known for its rows of glass greenhouses, and the Iwata Nursery had eight of them on a six-acre piece of land on a street lined with Wakayama carnation and chrysanthemum growers.

Sumi never knew her mother, Jiu, because she had died three days after giving birth to her. An older cousin raised Sumi, who was sick during most of her childhood. She had typhoid fever, in fact, and the whole house was quarantined. A special nurse was the only one allowed in the house with Sumi; her brother and two sisters had to speak to her through a closed window because they couldn't come in. She eventually recovered and went on to finish grammar school. Sumi was attending Mountain View High School when the war broke out.

The Iwatas, like our family, had moved to Colorado early and had not been placed into concentration camps. According to Sumi, a *Hakujin* florist named Mr. Foster had been influential at the time.

She remembers:

> *He told my father, 'There are too many girls in your family, you better not go to the camps.' He lived in Denver for a while and he said that it was a very nice place. So I guess he talked to my father and my father sent my brother to Denver. He bought a house and he came back. And then there's no questions asked. We just threw things on the truck and then my sister drove the other car and we left. My brother's best friend moved into the house [in Mountain View]. The nursery was left with no one to care for it. Later an Italian family managed the nursery.*

I had, in fact, stopped by the Iwata's house in Denver when I reported for my induction physical during the war. I discovered at that time that Maki had gotten married. But I had still not gotten a chance to meet Sumi. Not until that JACL dinner in Santa Clara, three years later.

After the JACL function, I began to stop by the Iwata house in Mountain View to see Sumi whenever I had to make a delivery of cucumbers to a pickling cannery, usually about two times a week. I'd also pick Sumi up when I bowled in a league in Mountain View. Sometimes we would triple-date with my close friends Hiromi Nagareda and Bill Kuwada. I spent so much time with those two that we called ourselves the Three Musketeers.

I finally married Sumi the following year, January 1947. At that time, not too many Nisei were marrying because we were all struggling to rebuild our lives. Our wedding ceremony was held at the San Jose Buddhist Church, an impressive building originally constructed by the Nishiura brothers. We had the reception at a Chinese restaurant called Fujinoya. The whole dinner, for about eighty people, cost $125. Sumi's dad insisted that he pay for it all. "I'm not selling my daughter," he had said when my dad had gone over there with some money for the wedding. Sumi even paid for her own wedding cake; it cost her a grand total of $25.

We lived in a little white house that I had fixed up on the ranch. One advantage of getting married, according to Sumi, was that we were eligible to buy rare goods like appliances and bed linens. It was hard to get those things, even after the war. I got a living room set, refrigerator, and stove, while Sumi bought a bedroom set, sewing machine, vacuum cleaner, and washing machine. Since we lived on butane on the ranch, I had to convert the gas washing machine to electricity.

Our lives were simple during those early years. We shared one car and didn't have a telephone for a while, so we had to go into town for our messages. It wasn't easy at times. But since then and up to now, we've always managed to get along. We go good together. Sumi says that our personalities are as different as night and day: she does things fast, while I like to think before leaping. But whenever I felt like I needed to take a gamble, she'd be there in my corner to support me. That's really something you can't take for granted.

CHAPTER

14

FOUNDING THE
HOKUBEI MAINICHI

While my father got back into farming, he was also called into community service. He became active with the Buddhist Churches of America (BCA) in San Francisco. In fact, when Bishop Otani, an esteemed leader within the Buddhist church, came to the United States from Japan for the first time, it was my dad who drove him around the West Coast in his car. Church work, as it turned out, not only involved attending dinners and driving dignitaries, but also, strangely enough, financing a newspaper. Before the war, the Japanese community in San Francisco had several daily newspapers. One was the *Nichi Bei Shimbun* (Japanese American News), which had been created out of two newspapers in 1899. One of its founders was Kyutaro Abiko, a Christian convert and labor contractor who helped establish the foundation of a Christian community called the Yamato Colony in Livingston, California. Another daily newspaper was *Shin Sekai Asahi* (New World Sun), which itself had been a result of a merger between *Shin Sekai Shimbun* (New World News) and the *Hokubei Asahi* (North American Sun).

Before the war, I had been too young to be interested in reading those kinds of newspapers, even though our family had some personal ties to Mr. Abiko, who used to come to the house. My father and Mr. Akibo, in fact, arranged the marriage between our widowed family friend, Mrs. Kami, and her second husband, a reporter for the newspaper's Sacramento bureau.

Sacramento also had its own Japanese newspaper, the *Ofu Nippo*. In 1939 a printer named Shigeki Oka took over the paper and continued printing it until World War II. According to his daughter Michi Onuma, Oka had been a "rather rambunctious person" who had originally worked at a relative's newspaper in Tokyo at the turn of the twentieth century. After getting into a fight with one of the editors, he was sent to the United States for study. Oka eventually became involved in seasonal work before launching his own print shop, Kinmon (which means "Golden Gate") Printing, in San Francisco. He specialized in bilingual Japanese-English catalogs, business cards, and flyers—even producing a war propaganda pamphlet for the consul general of Great Britain in San Francisco before the war broke out. This job led to an even greater mission: Oka and his lead type from Sacramento were sent to India and Burma to produce anti-Japanese leaflets for the British government. Oka served overseas for two years, eventually rejoining his family in Denver in 1945. The family eventually made their way back west to San Francisco, where Oka again attempted to create another newspaper, *Shin Shin Shimbun* (Progressive News), in 1946.

That same year, Yasuo Abiko, the son of the former *Nichi Bei Shimbun* publisher, and five other men started a new paper in San Francisco. They called it the *Nichi Bei Times* (Japanese American Times). Since the founders were more oriented toward Christianity, the Buddhists felt that they too needed a newspaper of their own. The three most committed to this venture were head of the BCA Ryotei Matsukage, publisher of the BCA newsletter Joshin Motoyoshi, and community leader Sasato Yamate.

This group then came to my father and other leaders within the BCA for financial support. I'm not sure exactly how much money my father gave, but he ended up with at least 20 percent of the company. According to newspaper historians, it initially had $30,000 in starting capital.

The group purchased the lead type from Oka's failing *Shin Shin Shimbun* and created a daily newspaper like the *Nichibei*, with separate pages in both Japanese and English. They named it the *Hokubei*

Mainichi (North American Daily) and published the first issue on 18 February 1948 from an office at 1737 Sutter Street in San Francisco's Japantown. Oka's daughter Michi Onuma had learned how to work a linotype machine in Denver, so she became the *Hokubei Mainichi*'s first English-language editor.

My father's friend, Kakuzo Ishimaru, another San Jose farmer who was definitely ahead of his time, became the first president of the *Hokubei Mainichi* in 1948. The following year, my father became the second, and stayed president until his death. Even though the *Nichi Bei* was supposedly the Christian paper and the *Hokubei*, the Buddhist, there's really no evidence of that today in terms of content. For example, Michi Onuma, who had worked on the *Hokubei*'s first English-language issues, later became connected with the competing *Nichi Bei*.

Today, most cities have become one-newspaper towns, but in San Francisco, both Japanese dailies are still in operation. That really says something about their Northern and Central California subscribers. I'm not active with the newspaper world myself, but the *Hokubei Mainichi* regularly comes in the mail to our Camarillo house.

CHAPTER

15

Our Farms

Meanwhile, in the late 1940s, Hirasaki Farms, Inc., continued to grow. We had grown from six hundred acres to approximately twelve hundred acres, in addition to our packing shed and shipping operation. The ranch had a good year in tomatoes around 1946, and we followed that up with lettuce, fall peas, and then celery.

During this time, we had to come up with some brand names and crate labels. This is where Schwabacher-Frey came in. Schwabacher-Frey was one of the largest commercial printers in San Francisco. One of their salesmen came around to farms in the late forties for business. The company would either design labels for existing brand names or else investigate possible names for farmers. If a name was not in conflict with anybody else's, Schwabacher-Frey could register it for you.

We already had come up with a name, Our Farms, because all of us siblings controlled everything. In addition, all of our product was grown on our own ranch. Our Farms would become our top label for our best produce. In the case of lettuce and fall peas, they had to be extra fresh and firm.

Lettuce was the earliest vegetable to come out in the spring. We didn't see celery until about June, and then by July, the hottest month, we'd be preparing the soil for fall peas and fall celery. The fields would be disced at least several times before a machine was brought in to create irrigation ditches. After spending a month in a hothouse in Venice, California, the celery plants would be delivered to Gilroy for planting in the fields. It was important to get the water running through the ditches within minutes of the initial planting.

Our second label was Country Pride. We chose names that weren't too long: two, three syllables were all we needed. They were easier to remember that way. We used Country Pride for produce that went to separate locales because different brokers would want different labels. One wouldn't want to touch the same label as another shipper or grower.

Our third label was Jimmy's Choice, named after my father, of course. This was for more of the local market, because everyone knew who Jimmy Hirasaki was. Our final label was H-F California Vegetables.

The packing process was key. You can avoid bruising if you have good packers. In the case of celery, the trimmed stalks would go through a pressure washer on a conveyer belt to the packers, who would box the celery according to size. You couldn't random pack. Most popular were the packs of two dozen, two and a half dozen, and finally three dozen per wooden crate. The celery in the three-dozen packs tended to be small. There were also the odd-sized dozen-and-a-half packs, or eighteens, which contained celery that were too big and hard to sell since consumers preferred medium-sized stalks.

Once the celery was packed, it would go through a large washer cooler, which is basically a big box in which the water percolates through the celery. It would also pre-cool the celery, getting the field heat off, before they were loaded into the railroad cars.

Since we were shipping all over California and even out of state, the sales managers had a large role to play. In the early days, we had Ruth Catherwood and Jack Nelson in our sales department. Later, a Chicago transplant named Denny Donovan Sr. joined the firm as a sales agent and my father's right-hand man. Born in Indiana and raised in Alabama, Donovan, in fact, was a good friend of Bill Crowley, a salesman for a label company. Donovan was a year younger than my dad. An outgoing man with bushy eyebrows and greying hair, Donovan had come to San Jose in 1945 to start his own produce brokerage firm. He had already cut his teeth as a buyer for West-Co, the purchasing arm of the Kroger retail chain in Chicago, and he could rely on his contacts there to sell produce from California.

I'm not sure how my dad and Donovan got together, but they had certainly traveled in the same business circles. Donovan's son, Denny Jr., didn't have the impression that my father was an ordinary farmer. Usually Dad would be dressed up, in pressed pants and a clean button-down shirt, and he felt very comfortable speaking English, unlike many other Issei. During the early 1950s, Dad and Denny produced a glossy black-and-white promotional booklet on Hirasaki Farms' celery operation. The spiral-bound booklet featured photographs of the ground being tilled, plants planted and irrigated, and celery harvested and cleaned. A line of women and young men, predominately Japanese American, can be seen packing the celery for shipment. A famous San Francisco head waiter, "Adolf of the Palace," was even photographed serving our fresh celery to patrons of the Palace Hotel.

Even though things were going well, I was itching for a change. Sumi had given birth to our first child and only son, Mark, in December 1948. I was a father myself now and although I wasn't aware of it at the time, I was considering the first of many new ventures that I would undertake in my life.

CHAPTER

INDEPENDENCE

It was in 1950 that I told my father I wanted to leave the family operation. My dad, on the other hand, wanted me to take his place. "Why don't you keep farming here?" he said. "Then you can take care of the family."

But I was twenty-seven years old with a wife and young son, and I thought that I could make my living doing anything. The worst thing that can happen to a son is to have everybody say to him, "You're lucky. Your father and mother are going to leave everything to you." That always left a sour taste in my mouth. I would say, "Well, I don't want it then." I figured that if I became a success, people would say, "The father and mother did it." And if I somehow lost the ranch, people would say, "He didn't know how to take care of it."

It was important for me to be independent, and I saw that it was a good time to make a break. The ranch was already paid for, and that was unusual in the produce business, in which everyone takes a high gamble each year. Most people end up borrowing lots of money and then spend the next few years or even decades fighting to pay it off. In contrast, my father, at fifty years of age, could retire. But he was still disappointed with my decision to leave, I know.

"It's too bad you think that way," he said. But in the end, he did let me go.

PART THREE: **SPRING**

Manabi Hirasaki stands proudly in the field of his Gilroy, California, farm, April 1959.

Gilroy Nisei 2 bowling team at 13th National Japanese American Citizens League Tournament. Manabi Hirasaki is pictured second from right. Los Angeles, 1959.

Sportsmen's Chef Club dinner, organized by Manabi Hirasaki. Issei women prepared Japanese food for this special community event. March 1961.

Z5-A strips, late 1950s. Strips were placed on the strawberry crates when shipped to market to mark the product's superior quality.

CHAPTER

New Start

Upon my decision to leave, Hirasaki Farms closed. For the next five years, I worked for Ghiselli Brothers, a large produce and shipping house in San Jose. They had purchased the celery packing and harvesting equipment from Hirasaki Farms and, while I was helping them put together our former equipment in their packing shed, they talked me into staying on. Paul Tarantino was the company's operator and Roy Barsotti was partner and accountant. They were strong men who provided me with a good training ground. In hindsight, I should have had an experience like that earlier. A son should always go to work for someone else before going back to a family business.

I was in charge of the packing crew, while another man, Ray Chu, oversaw the shipping side of the operation. The Ghiselli Brothers acreage was about three, four times larger than Hirasaki Farms and the production volume was a lot higher. They handled a variety of different crops. In fall, they would grow and pack endive and broccoli, but celery was their specialty.

As the head of the packing department, I learned how to take care of a crew. We had fifty to sixty people working as trimmers and packers, plus an overhead crew, in the shed at all times. Most of the workers were Issei and Nisei women, but we also had Japanese American men, as well as Mexicans and Filipinos, as part of our mixed crew. During the summertime, Nisei sons would come in and help out.

We exported more produce to overseas Army and Navy stations than anyone else in California. Since our crops had to travel far, we had to make sure that they were packed efficiently to prevent damage and decay. We usually packed in the evening. We used a hydro-cool process that would push the produce through a washer and then down a long belt. If the vegetables were a bit limp, the water would help expand and freshen them up. We would wrap the celery individually in wax paper and then crate them up according to size.

Refrigerated trains and trucks would take our produce to Navy transports docked in San Francisco and Oakland. We used an ice crusher that blew ice directly into the cars of stacked vegetable crates. There was also bunker icing in place to keep everything cool.

Dealing with the inside crew was a lot easier than overseeing the harvesting and growing. I had a subforeman who did a lot of the hiring and firing, but I had to get involved at times too, even though I was young enough to be the son of a lot of the workers. If someone wasn't doing the work, I had to let him or her go. I never fired anyone without reason, and the boss always supported me on my decisions.

I stayed at the company for five years. I tell people that things come in fives for me. I had stayed at Hirasaki Farms for about that length of time, and now I was ready to leave Ghiselli Brothers after five years. During this time, Sumi and I had another child, this time a daughter, Marcia Haru, in 1953. There were now four of us in the family and again, I was itching for a change. Little did I know that the change would come in the form of strawberries.

CHAPTER

STRAWBERRY DEAL

I first experienced the strawberry business when I was fourteen years old. We had subcontractors in Gilroy who did strawberries and raspberries for a short while. It was my job to pick up all the crates in the field and bring them underneath the shed in the shade where it was cool. I would help load the crates on the delivery truck and tag them with names of commission houses in San Francisco and Oakland.

But that was the extent of my early dealings with strawberries. Until the spring of 1955, I had been strictly a truck-farming man. Then one Friday afternoon a hometown lawyer and neighbor, Robert "Bob" K. Byers, came looking for me at home. He said he knew of a strawberry patch that was on the market. The field—thirty-five acres in Gilroy—was ready to pick. The original grower, Isao Ogawa, and his partner were contemplating a move down to Salinas before harvest time to develop a new ranch, and the processing company, Pic-Sweet, was looking for someone to take over. "Let's take a look," I told Bob.

Bob was about four, five years older than me. He had gone to Harvard and then UC Berkeley for law school. In the 1950s he joined his father's general law practice back in Gilroy. Like me, he was also looking for ways to better himself financially, so this strawberry deal seemed to come up at just the right time.

It rained over the weekend, but by Sunday evening the plants were still holding strong. A brisk breeze blew through the fields—perfect for drying the berries. "Let's buy it," we decided. We weren't going buy the land but instead buy the lease to farm the land for $60,000. But how were we going to finance it? Those days, strawberry growers raised their crops mostly for processing companies. So we went to the processor and borrowed the down payment—$30,000—from them. We also took out a $30,000 note from the owners that had to be paid in ninety days.

On Monday we went to the American Trust Bank and borrowed the harvesting money, an additional $15,000. The following day, I was out in the fields with the harvesting crew. So, in a matter of three days, Bob and I had gone into the strawberry business together without investing a penny of our own money.

At the time, it paid to be a strawberry grower. Strawberries were the thing to do. The dollar volume was very high: in other words, you received a lot for your investment. Everyone was willing to provide financing because they thought this kind of farming was a gold mine. Shippers, processors, bankers, brokerage houses, even people off the street would give you money to get into the berry game.

During the war, strawberry growing in the area had been at an all-time low. The industry was at a virtual standstill as we Japanese Americans were excluded from all of California—aside from the two concentration camps in Manzanar and Tule Lake—as well as from most of Oregon and Washington. Before the war, 90 percent of the strawberry crop was grown by Japanese sharecroppers, according to Miriam Wells's book, *Strawberry Fields: Politics, Class and Work in California Agriculture*. Strawberries were labor-intensive and called for the detailed handling that the Japanese were known for. Now that the specialized labor source was gone and the wartime demand called for vegetables, there was little production of strawberries.

Edward "Ned" Driscoll was one of the few who was still in the berry game during the war. He grew some acreage in Salinas for processors and also dabbled in his own experimentation. In 1945, there were only about 2,000 acres devoted to strawberries in

California. Ten years later, about the time Bob and I bought our strawberry patch, the state's total acreage had jumped to 22,500. A year later, according to George M. Darrow's classic book, *The Strawberry*, California was producing 55 percent of the entire nation's strawberry crop.

A lot of things contributed to this surge of strawberry growers. First of all, the Alien Land Laws, which had taken on different forms in 1913 and the 1920s, were finally declared unconstitutional by the California Supreme Court in 1952. Also that year, a legislative bill was established so that anyone, even Japanese, could become naturalized. Now the Issei could freely buy land, become citizens, and establish credit and trust. Now large farmers were helping their longtime Japanese sharecroppers to become commercial growers.

Households were also hungry for strawberries. Everyone was looking for strawberries, both fresh and frozen ones. Other growers remember how much people loved frozen food back in the 1950s. Frozen food was such a novelty. Frigidaire refrigerators even had special frozen food compartments back then.

Even though I had never grown strawberries, I understood the basics of harvesting and scheduling. For instance, because we picked for a processor, we had to take the stems off the picked fruit. That wasn't hard; you just had to make sure that the berries were mature. You used your thumb to hold the stem down while removing the fruit.

Everything turned out so good, we were in hog heaven! We decided there was nothing like it. Since we were doing a good job, the processor asked us to take on additional acreage. We bought forty acres of former dairy land across the street and then another sixty acres of lowland, which required a lot of leveling and improvements. Things were looking good but, like everything else in farming, I knew we could only take it one season at a time.

CHAPTER

THE CRASH OF 1957

All farmers are gamblers. I saw it in my father. I see it in myself and other farmers. Sho Kobara, who grows both apples and strawberries on his twenty-three-acre farm in Watsonville, described it this way:

> *Farmers are gamblers. You got to fight the weather. You got to fight the bugs. You got to fight the price. And if you have nice good strawberries, it's not worth much. When you have junky berries, when everybody has problems, then it's worth more money. It's the complete opposite from retail in hardware. If you're going to buy a shoe, you pay more for good stuff. Farming, that's not the case. If everybody has a bad crop, you get a lot of money for it. If everybody has nice crops, you can't get much for it.*

When we started that sixty-acre Buena Vista ranch in Gilroy's low lands in the fall of 1957, it was the worst possible strawberry time. It was the worst in the sense that everyone was planting and producing berries. By the 1950s, Naturipe Berry Growers, which would become the largest strawberry cooperative in the nation, was going strong with mostly Japanese American growers. Naturipe's roots go back all the way to 1917, when the Central California Berry Growers Association was formed by both *Hakujin* and Issei producers. One of the original directors was Richard Francis "Dick" Driscoll. His sons, most notably Ned Driscoll, helped create the Strawberry Institute of California on November 16, 1944, and later Driscoll Strawberry Associates, Inc.

Smaller growers joined the Watsonville Berry Cooperative, organized by Masaru Ronald "Buzz" Noda, a Central California man who grew strawberries in Fresno after the war. In the San Joaquin Valley, strawberries were a May/June crop that followed two months of tomatoes. Watsonville, on the other hand, had fresh strawberries as early as April for three or four weeks. When it turned warm, growers harvested strawberries for the processor.

The year of 1957 was disastrous for everyone in Pajaro Valley and Santa Cruz County, price-wise and weather-wise. The prices were no good in the first place because of overproduction; then the bad weather hit us at the same time. It rained on May 20, right at the peak of harvesting. Of that crop, we must have lost three hundred to four hundred crates per acre, with each crate worth about $1.75 fresh. That loss made up 30 to 40 percent of our first crop on the Buena Vista ranch. Since we were growing for a processor, we had to take whatever price they offered.

Nevertheless, we were still in pretty good shape compared to other farmers who borrowed all they could to plant strawberries. Quite a number went bankrupt. Some had to take out additional loans to stay in the game. The strawberry business continued to slip in 1958 and 1959. Everyone was having a hard time.

But Bob and I stayed with Gilroy Berry Farms. In fact, we were able to make a financial turnaround back in 1960 and '61. By that time, we had another sixty acres in Gilroy. Since our processor had asked us to manage other strawberry ranches, our total acreage had climbed to approximately two hundred acres; at that time, eighty to one hundred acres were considered excessive. The biggest development, however, was that we were starting to sell fresh to a company that would prove to be instrumental to my strawberry future.

CHAPTER

The Banner, Shasta, and Lassen

If you get to talking to any California strawberry grower, he or she will eventually mention "university berries." University berries are varieties of strawberries developed by the University of California. Prior to the 1920s, most of the experimental breeding took place on private farms. The Banner strawberry variety, for example, was first grown commercially by R. F. Driscoll and Edward Reiter on a ranch in Watsonville in the early 1900s.

Reiter's grandson, Miles Reiter, is a present-day leader of the strawberry industry. Although he was not even born at the time of the Banner's creation, its legend has touched him:

> *This berry transformed California from supplying local markets to extending out to long-distance shipments. A very revolutionary berry.*

The Banner strawberries, which were known for their sweet taste and superior quality, were soon adopted by other growers. Once this particular variety, selected from seedlings or propagated from the original mother plant in Shasta County, made a name for itself, other growers soon realized that developing new varieties could help sell strawberries. Even more importantly, they realized that they needed new strawberries that could withstand disease and pests, namely yellows and spider mites, respectively.

Yellows was a new disease in the Pajaro Valley district in the 1920s. Also known by its scientific name, *xanthosis*, yellows is caused by a virus. According to strawberry expert Stephen Wilhelm, yellows can stunt the growth of the entire plant, while also curling and damaging the leaves. The most threatening aspect of yellows in the twenties was that it not only hit the second-generation plants but also had infested the first-generation plants in nurseries up in Shasta County. It had the potential to wipe out the entire strawberry industry in California.

To combat such diseases, the University of California at Davis began breeding strawberries in 1925. The key prewar pathologist was Dr. Harold E. Thomas, and Earl V. Goldsmith was an important breeder. I didn't know either one of them well, but those who worked with Dr. Thomas described him as a disciplined, private academic who was dedicated to solving problems regarding disease control. Goldsmith, on the other hand, was not formally trained as a scientist. In fact, he did not even complete high school, but he had an even better education right on the farm. Even more importantly, he had a true passion for the strawberry. According to a U.S. Department of Agriculture report, Goldsmith was "a rather small man, full of energy, a keen observer, and an individualist. He was an idealist and devoted his entire energy to his strawberry work—to breeding ideal varieties." Goldsmith first worked as a ranch foreman for a prune grower in San Jose before being hired by Dr. Thomas as a foreman of the University of California Deciduous Fruit Field Station in Santa Clara Valley. He was in charge of field trials before he started crossing and raising seedling strawberries on his own.

Goldsmith wasn't the only nonacademic to be interested in the breeding of strawberries. R. F. Driscoll's son Ned was primarily a grower, but he had also actively involved himself in experimentation. He had test plots on his California farms as early as the 1930s. Harold Johnson, a former breeder for Driscoll Strawberry Associates (DSA), describes Ned's achievements:

> *He was ahead of most everybody. And that's why this whole concept of a private grower*

subsidizing research is unique. Everybody else in most crops depends upon the University to do the research for them and they adapt that. But for a grower himself to finance private research...it's done now more by other crops, but at that time, it was quite unique. And it was very, very important. Even in the midst of these stressful times, Ned saw the value of that, because he saw that if you didn't have a variety, you just didn't have it.

Ned loved strawberries so much that he almost went broke several times in search of the best variety. His friends say that he was a long-range thinker. He could talk about strawberries for hours. One of those who also shared Ned's passion was a Nisei man named Kay Mukai. He was actually the son of one of Ned's sharecroppers on a ranch in San Juan Bautista, California. It was Kay who took care of two test plots on Driscoll land for the University of California in 1937. His work would lead to the selection of two new university varieties, the Shasta and the Lassen, which would produce a fortune for the California strawberry industry. Both the Shasta and the Lassen were widely grown throughout the state for the next twenty years.

On my ranches in Gilroy, I grew all university berries—mostly Shastas and a few Lassens. I found the Shasta, a favorite of processors, to be sweeter, while the Lassen, more appropriate for a Southern California climate, was pinkish and a bit watery. What was most desirable for the processing company was the color. They wanted a deep red, solid type of meat, so the Shasta was ideal.

In a sense, the Shasta and Lassen were developed through close cooperation of the university and private growers. But the strawberry breeding world was just beginning to change in the 1950s. In 1944 Ned Driscoll had launched his own private research company, the Strawberry Institute of California, in Morgan Hill, and his first hires were none other than experts formerly with the University of California: Dr. Harold Thomas and Earl Goldsmith.

CHAPTER

NED DRISCOLL AND THE Z5-A

I had had some first-hand experience with Ned Driscoll during my first five years as a strawberry farmer in Gilroy. One of my ranches was right next to his. We didn't talk much at that time, and when we did, he teased me about being a "windshield farmer," which meant I was usually driving around, overseeing different ranches. In the late 1950s and '60s, I was often leapfrogging and starting new farms.

Ned was on the short side physically, but a big man in spirit. Those who knew him well describe him as enjoying people who had a desire for excellence. He had little patience with people who were sloppy or had a low level of commitment. He could be gruff and outspoken at times, but I was able to get along with him.

In 1956 I began selling some of my university strawberries fresh for the Driscoll organization. At that time, our company, Gilroy Berry Farms, was the largest independent strawberry grower in Gilroy. I wasn't an official Driscoll grower, so I couldn't handle special Driscoll patented varieties, which usually came out at the end of May and June, but I instead grew Shastas, which peaked in April and May. Driscoll needed an early berry, so one of their salesmen, Bill Crowley, approached me to provide fresh berries that would be distributed by DSA. Crowley was a good friend of my father's former right-hand man, Denny Donovan Sr., and he used to sell packing labels to Hirasaki Farms. He was also a Gilroy hometown boy. With all these connections, I began to sell to Driscoll's for a couple months out of the year in early spring.

While I was concentrating on Shastas, Driscoll came out with a groundbreaking berry in 1958. It was called the Z5-A, or sometimes the Goldsmith, but most folks knew it as the Z5-A. The Z5-A really established Driscoll as being the top breeder of strawberries in the world as well as a leader in establishing quality standards. Harold Johnson remembers this about Z5-A:

> *Even though it had very poor vigor, it was quite a spectacular looking fruit. It just had a lot of appeal. It gave us a vision of what could happen to get that continuous production.*

The Z5-A came out later than the Shastas, in mid-summer. As described in Darrow's book *The Strawberry*, it was large, with high gloss and "remarkable carrying quality." It was the Z5-A that convinced the Uyematsu family of Watsonville, former sharecroppers, to partner with their former landlord, Tom Porter, and grow Driscoll strawberries after World War II. The Z5-A was so popular in fact that it was patented in the late 1950s. As a result, only those who were Driscoll growers or paid a licensing fee could grow Z5-A strawberries. It was only a matter of time until the name "Z5-A" was featured on strawberry crates and sold to grocery stores as a special brand. Some grocery stores even began to arrange the Z5-As in baskets with their conical tips up so consumers could identify them immediately.

In the early 1960s, I expanded into Watsonville onto, I learned later, the same piece of land on which the original Banner variety had been first planted by the Driscoll and Reiter families. Looking back, I should have known that this was a sign of what lay ahead.

CHAPTER

WATSONVILLE

From Gilroy, you can get to the town of Watsonville two ways. You can backtrack south down toward Salinas and take Highway 129 west. Or you can go through Hecker Pass. Hecker Pass is a winding highway that goes through green hills, brush, and a line of vineyards. It was originally called Bodfish Canyon Road, according to Watsonville historian Betty Lewis, after a man named Bodfish who cut and appropriated timber that was under a land grant called Salsipuedes. Apparently he charmed the angry Salsipuedes ranchers by feeding them chicken, turkey, and iced champagne. The road was changed from Bodfish Canyon Road to Hecker Pass Highway in 1928.

Once you get to downtown Watsonville, you'll hit Main Street. From there, you will eventually reach a large bridge. It used to be, before the war, that the south side of that bridge was Chinatown. That's where the early Japanese pioneers settled after immigrant seasonal laborers first came to Pajaro Valley in 1892.

Charles Iwami's father, Yasutaro, came to Watsonville from Yamaguchi Prefecture around 1900 after stops in Victoria, Washington, and in San Francisco. Since Watsonville was such a small town, Japanese from different prefectures mixed together. Eventually the Japanese started another Japantown on the north side of the bridge. Yasutaro Iwami opened a barbershop on Main Street, alongside pool halls, grocery stores, laundries, tailors, and a tofu store.

The Pajaro River brought both joy and terror. Charles Iwami's wife, Nancy, said it looked like the Old West, with trees and trails on each side of the river. The boys went fishing, catching suckers, carp, chubs, pike, and river perch. The fish weren't for eating but for placing in fish ponds in people's yards. In fact, the only time I crossed over to Watsonville as a kid was to buy lures from a fishing tackle store famous for its handmade fishing poles.

There was no levee at the time, so two times a year the Pajaro Valley would flood during the rainy season. The Iwami's home was located in back of the barbershop storefront, down a hallway that also adjoined the pool hall room, kitchen, and indoor Japanese bath. Bedrooms were located upstairs. In the months of December, January, and February, the Iwami family began moving sand bags to the front of the building and stuffing long, thick strips of dried seaweed called *kobu* in between the doorjambs and the spaces between the doors and the floor. The piano was moved on top of a large wooden *usu*, a giant mortar used to pound *mochi*, Japanese rice cakes. The beds were placed on top of wooden boxes. Then the water would come, seeping in the downstairs rooms for three or four days. Sightseers would float past in boats.

The Iwamis tried to stay upstairs as much as possible, but when they had to go downstairs, they wore high boots. After the waters receded, mud would be left on the linoleum and wooden floors, leaving a big clean-up job for the Iwamis, who had to make sure that the barbershop area was as clean as possible.

The Issei first farmed sugar beets in Watsonville but soon moved into strawberries. As Kazuko Nakane says in her book, *Nothing Left in My Hands*, cultivating strawberries was easier than hoeing sugar beets. And after labor disputes with the large sugar company Spreckels, Japanese laborers opted for contract leases for strawberry farms in which equipment and other supplies were all provided. In time, Issei pioneers, including Unosuke Shikuma and Harry K. Sakata, were forming partnerships and corporations to purchase their own land.

I experienced Watsonville much later than this time, in the early 1960s. I first leased thirty-five acres of a La Selva Beach ranch and then fifty acres of the Cassin ranch located in the West Hills. Even though Gilroy was right next to Watsonville, they were two different worlds. I never really became a part of Watsonville. There were the Christians and there were the Buddhists; the Japanese community was pretty much divided along those lines. Before the war, there were at least four hundred people living within city limits. Many never returned to the area, but the ones who did most likely turned once again to farming to make a living.

Whereas Gilroy was hot and dry, Watsonville held more of the cool ocean air from the Monterey Bay area. The ground was heavy and the wind was strong. I remember that I kept losing our tarp on the fields around 1961 and 1962. I still kept farming under our independent company, Gilroy Berry Farms, and kept up being a "windshield farmer," driving from one ranch to another. This was after my fifth year of farming strawberries. The changes in labor laws and policies finally convinced me that I needed to make another change in my life.

CHAPTER

BRACEROS

Labor had changed quite a bit since my father used subcontractors on his farm in Gilroy. When I returned to farming on my own in the 1950s, it was all about *braceros*, literally "ones who use their arms" in Spanish. *Braceros* were laborers brought in from Mexico to work on farms in the United States for a limited time. Created through an agreement between the Mexican and U.S. governments in 1942, the *bracero* program provided a steady labor force in California while many Japanese Americans had been removed from the West Coast and men of all ethnicities went off to war or to work in war-related industries.

Even when World War II ended, the *bracero* program continued. *Braceros* were lifesavers to us strawberry growers. Strawberries are a labor-intensive crop that requires human hands to do the picking. Even to this day, you can't rely on machines to deal with this delicate crop. So when Bob Byers and I had to harvest that first strawberry field in Gilroy, we turned to the *braceros*. We went through an association that had connections with the U.S. Labor Department and the Mexican government. They had a labor camp in El Centro, California, near the U.S.-Mexico border in Imperial Valley. Our recruiters went down to the labor camp to pick up workers. First the workers were fed one or two meals while they were getting processed and then we had buses for them to travel up to Gilroy. We had to guarantee at least four hours of work a day for at least six to eight weeks at minimum wage. Their room and board, provided by the labor association, was deducted from their paychecks.

We did our own payroll, but we had to pay a contracting fee to the association. Before I got into the berry game, 10 percent of the *braceros*' pay from 1942 to 1949 was to be deducted and transferred through the U.S. government to Mexican banks as a special incentive savings plan. Once the *braceros* returned back to Mexico, that money was to be spent for farm equipment. Apparently, many never received the balance of their pay, and as a result, class action lawsuits have been filed as recently as 2001, seeking a return of the money. The current Mexican government has also shown an interest in investigating this issue. Since I was involved with the *bracero* program after 1949, I don't know much about what happened with that deal.

Most Japanese American growers say they had good experiences with *braceros*. We depended on a good labor source, after all. Mexican locals served as foremen to train the newly arrived workers. We would usually bring them in slowly. Usually I had three *braceros* per acre during the peak harvesting period. So on a fifty-acre ranch, you would have to prepare for one hundred and fifty *braceros* in total. If I knew that my crop was starting to bloom on April 15, I'd bring fifteen to twenty people in. A few days later, I'd bring forty to fifty people, and so on. A lot of the time, certain farmers would be short or long on workers, so then we would negotiate a trade. You have to go by rule of thumb with everything.

The *bracero* program would finally come to an end in the early 1960s. Although it was tough on a small commercial farmer like myself, it probably was a good thing. If the program continued, we wouldn't be able to compete with the larger operations, who could easily contract hundreds of *braceros* at any time. The biggest challenge of producing strawberries is to hire labor during the peak periods. You do this by carefully watching schedules and cultivating relationships. Now, with the *bracero* program curtailed, all of us farmers were in the same boat, looking for the same thing: how to find people to pick the strawberries.

CHAPTER

PROTECTING THE ROOTS

When I was in Gilroy, just coming into Watsonville, I had heard there had been potato in the ground of this sixty-acre piece of land I was working on. Even though that had been a long time ago, I should have known that was bad news. I learned how bad from a professor named Stephen Wilhelm.

Wilhelm was considered a pro in fumigation and disease control. He got his degrees in plant pathology from UCLA and UC Berkeley and was working for the university system when I met him back in the 1960s. He was the one who introduced strawberry producers to methyl bromide fumigation, a system that has a bad name today but was a lifesaver for us growers back then.

Methyl bromide is a manmade chemical injected into the soil under plastic mulch to sanitize the ground, keeping it free of viruses and other pathogens. Before methyl bromide we had had to move from one piece of virgin land to another; with it, we could stay on the same ranch year after year.

Without methyl bromide, you had to be especially careful with land that had once supported tomatoes and potatoes because those crops carry the disease verticillium wilt, which attacks the roots of the strawberry plant. A plant could start out growing normally, but once it produced fruit, verticillium wilt would strain it, keeping the roots from supplying enough water to the rest of the plant. It's especially bad in the summertime. Soon the whole plant just collapses; it dies and you're finished. Methyl bromide, used in combination with the chemical chloropicrin, is most effective in fighting the disease.

For Professor Wilhelm to test our soil for verticillium wilt, I had to square off the ranch and have it plotted so each small package of test soil I took to the university could be identified properly. I had to record exactly where I got the soil—including the depth—and each sample had to be numbered. He was very particular, and it was tedious work.

As it turned out, there was no hope for our sixty acres. Methyl bromide had yet to be fully developed, so there was nothing we could do to fight the verticillium wilt. We had to level the whole place and eventually abandon it.

Since then, the injection of methyl bromide has become an integral part of the preplanting stage. But it has its problems. After it was shown to deplete the ozone layer, the United Nations mandated a worldwide ban on its use. The ban has, however, been postponed at least twice because some countries say they need more time to comply. Some California strawberry growers have complained that it's not fair for U.S. companies to be required to stop while overseas competitors continue to use methyl bromide. Like everything else about farming, there are no easy answers.

CHAPTER

BROKEN BOTTLES

When Hirasaki Farms was still in operation, my father received a large delivery at the Gilroy house: a truckload of *shoyu*, soy sauce, from a company in Japan. Japanese companies were trying to get back on their feet after World War II, and some tried to rope my dad into helping in some way. But my father was a farmer, not a merchant, so things sometimes turned out badly. In the case of the *shoyu*, the product was contained in glass bottles. When my father shipped the bottles, they often arrived broken, and Dad was the one who ended up paying for all the losses.

After we closed the packing shed, he got more involved in Japanese enterprises. Each district in Japan has its own specialty, and my dad would work with different growers' associations to introduce new crops to the area. As a result, he lived in both Japan and Gilroy, later moving my ailing mother to Japan with him. He bought some garlic from Manchuria and tried to grow it in Japan. My dad also found some good pepper seeds and attempted to raise those in Oita Prefecture, a place right next door to my dad's hometown area of Kumamoto that was known for its chili peppers and *wasabi* (horseradish). Chili peppers are valued by their heat—the hotter, the better. But the problem with peppers is that, like strawberries, they remain pure through the planting of seedlings, not seeds. With seeds, you don't know quite what you are going to get.

Also, the garlic wasn't good enough for export because it couldn't compete with the quality of Californian goods.

One of my dad's last projects involved strawberries. He asked me to buy Shasta plants for a growers' group in Nagasaki, located on the southern end of Japan. In the sixties, the area was known for its capabilities to freeze fish, and then many businesses were also considering growing strawberries for the frozen food market. My father bought so many strawberry plants—enough to cover three hundred acres—that the Nagasaki customs people couldn't unload them from the boat when the shipment arrived. The authorities, moreover, were afraid that these imports would spread disease, so they stuck the plants in a warehouse, where they eventually rotted. My father was allowed to take a few seedlings out to plant for experimental purposes, but that was all.

At some point I think that my dad stopped telling me about his different projects. I remember once seeing a newspaper article describing how he was eventually able to ship ten tons of Japan-born strawberries into California. After he died, one of his obituaries mentioned that he had also been importing ice-cream cones from Japan. I knew nothing about that. He probably figured I would just comment, "More trouble."

CHAPTER

10

WILD GAME

I was mostly busy with work in the 1950s and '60s, but I did make some time to have fun. Sumi, on the other hand, spent most of her time raising the children and taking care of the household. During our early married life, she helped pack strawberries into crates, our family friend Hiroko Yamano looking after Mark and Marcia while we both were in the fields. Later Sumi also served as the bookkeeper and payroll manager for my ranches.

Once a week I bowled in the Gilroy Nisei League. You might not think there were enough Japanese Americans to make up a league, but Gilroy Bowl on Main Street only had ten lanes, so it wasn't a problem. I was just an average player, bowling about 170. I remember driving down to Los Angeles in my 1957 Chevy for a JACL bowling tournament at Holiday Bowl in the Crenshaw District. My friends and I were part of the Gilroy Nisei 2 team, so we weren't too spectacular, but it was nice to spend time with old friends.

I was also part of the Gilroy Sportsmen's Chef Club, a local men's dining group organized after the war that welcomed all races. We would meet about once a year and enjoy barbecues of fresh wild game that the members had hunted. We ate duck, deer, bear, and even snake. A local restaurateur, Bill Corsiglia, opened up his place in the south side of Gilroy for the events. In 1961 the group decided to have a Japanese dinner, and I was named committee chairman. I actually didn't do any cooking, but I recruited the local Issei ladies to help prepare the food. We had a feast of steamed lobster, rolls of *makizushi*, and other Japanese delicacies. Everyone, even the Japanese cooks, had a fun time. I didn't feel any discrimination—we were all hometown folk.

Those two groups plus the Kiwanis Club of Gilroy were the only social and community activities I was directly involved with. My dad, however, had been at the center of civic activity. In the 1950s he decided that he wanted to donate a corner of an eight-acre piece of property we owned in town for a Japanese community hall. We had purchased that land in the late 1940s for our packing shed. According to the provisions of a private loan, the land was placed in my name. Since my dad wanted to give about half an acre to the Gilroy Nisei Athletic Club, the most active Japanese American group in town at the time and the primary user of the future community hall, I went ahead and made the changes on the deed. Volunteers helped move our old farmhouse from the Pacheco Pass ranch to the property (located near Tenth and Alexander) to serve as the building for our community hall. There the postwar Gilroy Buddhist Church met, in addition to the Japanese American Citizens League and sometimes even athletic groups; the community hall enjoyed its heyday in the 1960s and '70s. Now, many of those who used to gather there have moved away, and the building is in serious disrepair. A few fires have been inadvertently set by homeless men seeking warmth there during the winter months. When you drive past the community hall today, you'd never know that it had at one time been the center of an active community.

CHAPTER

MOTHERS, DAUGHTERS, FATHERS, AND PREGNANT PLANTS

I used to always say that when the strawberry plants finally came to us from the cooler areas of Mount Shasta, they would arrive "pregnant." What I meant was that the bloom was already starting to show. Pretty soon after I got those plants in the ground, I'd be seeing some flowers. In those days, however, you didn't get too much fruit in the first year. The second and third years would be harvesting time.

Actually, the plants we received would be the "daughters," the product of the original plants, the "mothers." Unlike onions or carrots, commercial strawberries don't come from seeds. If we used seeds, every plant would be a little different and thus less uniform. As a result, commercial strawberries are grown from runner plants, which are the additional growths that come from a mother plant. The ones saved for planting are shipped down to strawberry ranches in plastic bags or bins, dirt and all.

As I mentioned before, we used to hang onto a plant for two or three seasons. If our crop was thin, we would take the runners growing out of the plant and stick them into the soil farther down the row. These days you wouldn't do such a thing. After one season, all the plants would be dug up and replaced with new daughters.

Plant breeders experimented with the daughters of new varieties. Al Amorao, born in the Philippines and raised in San Martin, California, became involved in strawberries a little after I did. When he was in his early twenties, he got a job trimming strawberry plants for the Strawberry Institute of California, then located in San Martin. These plants were from cooler locations in Northern California, from places like MacArthur and Redding. A giant lawn mower cut the leaves and stems off, then a scoop dug up the crowns with the roots and dirt, and then the tumbler shook most of the dirt off. The plants were then placed in burlap sacks and sent to San Martin to be trimmed.

Al joined the trimming team in 1962. Back then he was known as Amado, but Dr. Harold Thomas said he was going to call him Al: "Even if I remember your name, I will never be able to pronounce it," he told Al. A year later Al joined the Air Force during the course of the Vietnam War. During his four-week furloughs, he returned to the Strawberry Institute for work. Harold Johnson, who had taken over research after Dr. Thomas retired, took him to Watsonville for the planting of test crops. Apparently something about Al stayed with Harold Johnson, because when Al was discharged in 1967, Harold offered him a job. Al recalls:

> *I had no intention in becoming a plant breeder...I was like any other person; I thought a strawberry is a strawberry. But once you become involved, you start seeing the difference.*

Harold Johnson had been hard at work developing the G3 variety. According to others in the strawberry industry, the G3 had both good taste and good size. It was finally named Heidi, after Harold's daughter who died in the mid-1960s when she was only thirteen years old. The Heidi would prove itself in later years.

We in the Hirasaki family also dealt with loss in the 1960s. First our mother, Haruye, who had suffered from diabetes and other health problems, died in Japan in July of 1960. My dad brought her ashes over to California and more than six hundred people attended her service at San Jose Buddhist Church. Then, on Christmas Eve of 1963, my father died at Gilroy Hospital. Three days later, we gathered again at San Jose Buddhist Church. We eight children now had no living parents at all.

I had been an independent grower for a little over five years. With both my parents gone, there was nothing to keep me in Gilroy, and it was time to move on.

CHAPTER

CULTURAL PRACTICES

In farming, we have what are called cultural practices. That means the way we handle plants, whether it be irrigating or fertilizing, it's not something you read in a manual, but what you experience out in the fields day after day. Watsonville grower Sho Kobara describes it as talking to his plants:

> *You don't physically talk to the plant. You actually walk through the fields and look at the color of the plant. You say, I guess it needs more fertilizer or water. But you need to do this in the fields. If you just observe from the road, the plants may look good, but you may miss seeing a certain kind of disease or problem inside of the plant. So you have to be constantly in the fields, nonchalantly pulling a few weeds or walking behind the workers. The plant talks to us by the way it looks. You start thinking, oh, maybe I'd better do this or maybe I'd better not do that. And sometimes you're wrong and sometimes you're right.*

Everyone had his or her particular way of doing things. For me, it was important to be a neat farmer. No weeds in the field. By keeping things clean, your diseases are fewer. Any ground that once was home to plants with a deep root system—potatoes and tomatoes, for example—usually carried disease.

In my day, the cyclamen mite and spider mite were big problems. The cyclamen mite is a tiny pinkish bug that lives in the folds of unopened leaves, deep in the crowns; the spider mite devours the undersides of the leaves. Since plants receive all their nutrients and moisture through these leaves, any deterioration causes the fruit to shrink up. The spider mites hid on the bottom of leaves, so if you sprayed insecticide only on top of the plant, the mites would still be around. In Watsonville we sprayed by hand. Hand spraying was a big job. You had to place large equipment in the fields, in between the furrows. A crew of ten men, each wearing rubber shoes, rubber coats, and face masks, pulled hoses through the rows of strawberry plants. At the end of each hose was a powerful spray gun with enough pressure to blow the leaves upside down to get to the spider mites resting underneath the leaves. In flat areas like Gilroy you could machine spray two rows of berries at one time.

For cyclamen mites, we'd place plastic tarps over the strawberry beds and inject methyl bromide underneath the plastic. In the early days, the ground would get so hot underneath the tarp that the heat would damage our plants and we wouldn't be able to do anything for about three weeks in between crops. Now, we'll fumigate the soil for diseases and weeds even before we'll plant. This is called the preplanting stage. A couple of folks at Driscoll's even created a giant insect vacuum cleaner called the BugVac. Others implemented organic farming practices, ordering and releasing bugs that would prey upon the pests that would damage strawberries.

Irrigation was also important. Before the 1960s, we nailed three long wooden boards together to form a "U" shape. These would become the wooden flumes through which we'd irrigate the strawberry beds. At the head of each row we had cork plugs, which we'd open until each bed was fully watered. Sometimes on hot days we'd even irrigate an extra time after picking. On gravel ground, in which the water would just seep through the ground, we had to irrigate two times.

These methods were all part of cultural practices, learned by watching other people. You follow a general program. If you're planting tomatoes, you know that you're supposed to be planting in March. You don't want to be planting in the summertime or in the middle of winter when the fog can come and get you. You follow others' examples but then again you also use your own timing on it. Timing is very important. Let's say it's spring and you're planting strawberries. You notice that it looks overcast; it's going to rain next week. You want to try to either beat it or do it after the rain. You follow your intuition on that. If you think that it is going to be a big rain, you know that you will be delayed quite a bit if you wait until it stops. So you hurry up and do the planting.

As Sho Kobara says, every year is different. Swede Johnson, one of the partners in Driscoll, would tell him, "Anybody can grow strawberries in the spring, but it takes a good farmer to have strawberries in the fall." It's true—you have to keep doing things differently depending on the time of the year. That's when it goes back to walking the fields and talking to the plants.

CHAPTER

HEADING SOUTH

By 1964 I was planning to quit strawberries for good. Since the *bracero* program had ended, I was having a hard time getting laborers. We had recruiters go to Globe, New Mexico, and El Centro, California. From New Mexico we got a group of Native Americans who were completely unfamiliar with strawberry growing and picking. Some labor contractors went into Los Angeles' skid row and put homeless men—some even intoxicated—on buses to work our farms. After breakfast, many of the contract laborers were gone.

A number of the larger growers went out of business in the early 1960s. It was too difficult to find strawberry harvesters. Since I too had quite a bit of overhead, I worked as a foreman for a receiving station at a Stokley tomato cannery in Gilroy in the fall, when there wasn't too much to do on the farm. That's when I first met tomato inspector John German.

John was tall, with thick brown hair and a big personality. When I first met him on that Stokley loading dock, I noticed that his leg was in a brace, but at no time did I think to ask what had happened to him. I felt that was his business and not mine. Decades later I learned that he had broken his neck in a car accident while he served in French Morocco with the National Guard. Through physical therapy, he was able to regain movement in his arms and lower body.

John and I got along from the very start. He was a fair man. We both had strong feelings about the right and wrong ways to do things. While I was quiet about my feelings, John didn't hold back. We always found things to laugh about; John says having a sense of humor is key in our business.

I only did the tomato work for one season. Like when I left my father's farm, I wasn't quite sure what I was going to do next. But I've always known when to leave. I guess I've always had good intuition in that regard. Then this new deal came up with one of Ned Driscoll's sons, Nod. He had a partnership in San Diego: three hundred acres of strawberries, all planted near the border with Mexico. There was a lot of farming in that general area, but few strawberry ranches. The soil was salty and, even with a large pool of labor, it was difficult to find pickers who would work five days a week. I was one of two supervisors and I knew that it wasn't the best arrangement.

After three months I returned to Gilroy. Then another new venture came along, again with Nod Driscoll, in Oxnard, California. This was coastal farmland just about one hour north of Los Angeles. When it came to Southern California, every strawberry grower thought of Orange County, specifically Anaheim. Anaheim formed one third of the strawberry-growing region in California, along with Salinas and Watsonville. Growers were interested in Southern California because the warmer weather allowed them to produce strawberries earlier in the season. Whereas Anaheim, with Disneyland and expanding suburbs, was getting overdeveloped, Oxnard had a lot of open land. But strawberries had never taken hold in Oxnard. Many, including strawberry pioneers like Joe Reiter and Cy Kennedy, had worked the land in the 1950s and '60s. I was going down to Oxnard with Nod Driscoll, and Driscoll's had no proven varieties for that region.

Being a Gilroy hometown boy, I had no direct connections with anyone in Oxnard. My wife, Sumi, and my two children—Mark, a senior in high school, and Marcia, an eighth-grader—would have to uproot themselves from everything they had ever known. But I didn't think of the costs to me or them at the time. I felt that I could succeed in Oxnard, even though I wasn't quite sure how.

PART FOUR: **SUMMER**

Sumi and Manabi Hirasaki dressed in farm attire at the Japanese American National Museum's Spring Festival. Los Angeles, 1993.

TOP: *Japanese American National Museum's Annual Dinner at the Century Plaza Hotel. From left to right: Sumi Hirasaki, Mayor Richard Riordan, Manabi Hirasaki, Akio Morita (chairman of the Sony Corporation), Yoshiko Morita. Los Angeles, October 1993.* **BOTTOM:** *Manabi Hirasaki and The Honorable Norman Y. Mineta at the Japanese American National Museum's Annual Meeting at the Los Angeles Hilton and Towers, February 1991.*

TOP: *Manabi Hirasaki kneeling in his strawberry field at Manabi Farms, Inc. Oxnard, California, 1990.* **BOTTOM:** *Manabi Hirasaki working in his office, 2001.*

Christmas at Manabi and Sumi Hirasaki's house in Camarillo, California. From left to right: Steve Messinger, Marcia Messinger, Mark Hirasaki, Sumi Hirasaki, and Manabi Hirasaki. December 1997.

The Manabi & Sumi Hirasaki National Resource Center was named at the Japanese American National Museum in December 1999. Photograph by Marvin Rand.
FOLLOWING PAGE: *Manabi Hirasaki carrying a box of strawberries at Kita Farms. Oxnard, California, 1999.*

CHAPTER

Early Sign

When I moved down to Oxnard in July 1965, I had to start from scratch. We had lost the option on a previous ranch, so I had to find another one. When a farmer moves to a new territory, he has to work carefully. He needs to stand back, observe other ranches, and form relationships with other farmers. Nobody is going to help you if there's no direct connection. I had no one to give me any advice, so I was in a bad situation.

I had one friend, Mack Hamaguchi, who was a real estate man in Gardena, a suburb south of Los Angeles. Mack told me that he was on his way to Oxnard and asked me to meet him at a celery ranch operated by the Chikasawa family. I drove over to the ranch and waited for him there. As I looked around at the celery equipment and packing shed, I felt as though I was being reunited with an old friend. It was the exact same equipment I had used at Ghiselli Brothers, the same equipment from Hirasaki Farms! I had repaired that equipment by welding together certain parts, and this machinery had my mark all over it.

Since I had left the San Jose outfit in 1955, Ghiselli Brothers had quit the celery business and apparently sold the equipment to this Japanese American farmer in Oxnard. Even the field baskets had come from San Jose; they were stamped "Hirasaki Farms." This was an early sign that things were going to work out in Oxnard.

Later that day we went to visit a strawberry farmer named Jimmy Arimura. He told us about a piece of land next door to him that had just been sold by its previous owner, Mr. Pfeiler, to the Iba brothers in Orange County. It was forty acres (in those days, Oxnard was chopped up in forty-square-acre grids). The former landholder, Mr. Pfeiler, was staying on the land in a share-rent deal because the Iba brothers couldn't find another tenant at that time. With share rent, the landlord would receive 10 to 25 percent of the crop. As a result, we thought that the Iba brothers might be interested in leasing the land to us rather than continuing a share-rent agreement. I dashed over to the ranch. Chicken manure had been spread out to prepare the land for planting lima beans. Mr. Pfeiler was more than willing to leave the land, as long as we paid for the chicken manure and his other expenses.

So, in a matter of hours by telephone, we were set. We had found our first home ranch in Oxnard, California.

CHAPTER

OXNARD

Oxnard was named after a family, specifically three brothers from New York that started a sugar beet factory in the area called the American Beet Sugar Company. Most farmland in California once had sugar beets. People used to think that sugar beets grew only in sandy soil, but they also thrived in heavy soil. Oxnard's ground was loose; we farmers called it sandy loam, which meant sandy ground with some body to it. Sugar beets were easy to pull in sandy soil. It wasn't like the hard ground in Colorado, where my father and I worked as contract farmers in 1942.

According to community historian Yoshio Fukuyama, a San Francisco labor contractor brought the first group of Japanese men to Oxnard in 1898, the same year the city was established. San Buenaventura, or Ventura, located just north of Oxnard, was founded more than a century earlier in 1782 for the development of Mission San Buenaventura. By the early 1900s, Chinese immigrants had settled in the downtown area of San Buenaventura; a Chinese man named Ung Hing had even purchased and lived for a short time in the historic Ortega Adobe located on Main Street.

Most Japanese made their homes in Oxnard to work on the sugar beet and citrus farms. When Oxnard was officially incorporated in 1903, a labor conflict was brewing between the workers and the largest labor contractor of the time, the Western Agricultural

Contracting Company (WACC). The Mexican and Japanese workers, in fact, got together to form the Japanese-Mexican Labor Association to protest their pay and working conditions. In addition to a large percentage of their wages going to the labor contractor, they were paid in script for high-priced merchandise from the company store, which was run by a Japanese American. The workers went on strike and WACC's power was finally broken.

One of the leaders of the labor association was a man named Kusaburo Baba, who had once served as a labor recruiter himself. Baba was a Kumamoto man like my dad. Unlike my dad, who was Buddhist, Baba was a strong Christian and helped to start the Methodist Mission Church, which he eventually pastored. He was against smoking, a stand which made him pretty unpopular among tobacco shop owners, and he was also opposed to gambling in town. Baba died in 1945 at the age of ninety-two, so I never knew him. I did know of another Baba in Oxnard who was as big as a *sumotori*. Since Kusaburo Baba had apparently been over six feet tall, I figured they must be related somehow.

Before World War II, most of the Japanese American businesses were concentrated on Oxnard Boulevard in between Fifth and Seventh Streets. They included Asahi Company, Inc., a general merchandise store that was incorporated in 1907 and eventually became Asahi Market, which was operated after the war by Oxnard native Nao Takasugi, one of the founder's sons. In fact, Takasugi was attempting to change the signs on his market when he met up with some resistance from city hall. His bad experience later led him to seek public office, and he served as mayor of Oxnard from 1982 to 1991 and later as a California state assemblyman from 1992 to 1996.

Even though Oxnard has had two Nisei mayors—Takasugi and dentist Tsujio Kato—the population of Japanese Americans in town has always been small. According to the U.S. Census, in 1940 there were only 672 Japanese in Ventura County, which includes Oxnard. Many of the old-timers had not returned to Oxnard after being forcibly removed in 1942. After World War II, Oxnard was full of transplants, or outsiders, coming in. A lot of strawberry growers went from Orange County to Oxnard, or vice versa. Some, like me, came from Watsonville. In that sense, many of us were strangers.

Some other postwar settlers were the Ito brothers, Kensaku and Yonejiro. Before the war, the Ito brothers' parents had farmed vegetables in Gardena, California. When the children went to Japan for their education in the late 1930s, they found themselves stranded there for a time. Making their way back to their homeland, the United States, the family began farming again in Garden Grove, Orange County, and then in neighboring Westminster. In 1961, Kensaku and Yonejiro expanded the family growing operation by moving up to Oxnard, while their other brother, Tomio, maintained the Orange County ranch. Yonejiro explains why they decided to expand beyond Orange County:

> *Smog used to just burn those celery and made them turn brownish. And whatever vegetables grew, it didn't look good, because of smog burn. Oxnard didn't have any problem with smog yet.*

Even though we were both growing strawberries, I didn't get to know the Itos for a long time. Everyone stayed in their own groups. I was part of Driscoll, while the Itos were members of the cooperative, Naturipe Berry Growers. Naturipe Berry Growers, like Driscoll Strawberry Associates, Inc., has its roots in the Central California Berry Growers Association in Watsonville. When the Central California Berry Growers Association reorganized in 1958, the name Naturipe Berry Growers was born. In the early days, Naturipe was made up of mostly small farms operated by Japanese Americans. The cooperative expanded into Orange County in 1963 due to the efforts of Tomio Ito and another pioneer, Paul Murata. The Muratas and my family were old-time friends, dating back before World War II. Paul's father was another Kumamoto man like my father. He initially specialized in chili peppers and even had a chili dehydrating plant on his ranch in Orange County.

About the same time that the cooperative strengthened its presence in Orange County, Naturipe went into Oxnard with the Ito brothers. While Driscoll Strawberry Associates today specializes in only fresh strawberries that it has patented, Naturipe handles both fresh and frozen university berries. The advantage of getting involved with a group like Naturipe or Driscoll is that you have access to large freezers and a centralized distribution network.

Technically speaking, I guess Naturipe and Driscoll were competitors, along with all of the many other strawberry cooperatives and companies, but I really didn't feel like I was in competition. We were all dealing with the same weather, pesticides, labor issues, and prices. We were all too busy trying to make a future in Oxnard that we didn't have time to worry about the next guy.

CHAPTER

E. F. DRISCOLL

FARMING TRUST

When I lived back in Gilroy, I went to the Milias Hotel on Main Street for the monthly Kiwanis Club meetings. The Milias, the oldest and largest hotel in Gilroy, served as a watering hole for most of the local growers. Ned Driscoll, for instance, liked to hang out at the hotel lounge in the late afternoon. Since I always tried to be early for meetings, I usually ended up socializing with Mr. Driscoll and his friends. I didn't drink alcohol like the rest of them, but satisfied my thirst with RC Cola and other soft drinks.

By this time, Mr. Driscoll's first wife had passed away and he was with his new wife, Bunny, who had been his assistant bookkeeper. The daughter of the last county game warden in the whole state of California, Bunny was a hometown girl and was about a decade ahead of me at Gilroy High School. I was younger than Ned's wife and older than Ned's children; you could say I was in between the generations.

Ned had suffered a stroke and was having a hard time. They needed someone to take over his privately operated ranches, which ran from Watsonville to Salinas to Santa Maria to Oxnard. About three hundred acres in all. One day Ned and Bunny came down to Oxnard to see me. It wasn't unusual for me to have dinner with them when they were in town. But this time they wanted to see if I would run Ned's farm for them as general manager of E. F. Driscoll Farming Trust.

This would be a big job that would require a lot of traveling and being away from home, but I could smell that this would be the opportunity of a lifetime. I accepted and from Tuesday to Friday I'd travel to Watsonville, Salinas, and Santa Maria, returning to Oxnard from Saturday to Monday. I did this for seven years.

Mr. Driscoll had once teased me about being a windshield farmer. And now, thanks to him and Bunny, I was the ultimate windshield farmer.

CHAPTER

Driscoll Strawberry Associates

If you are a fan of strawberries, you probably know the Driscoll name. From the 1990s, we began to place strawberries in clear plastic containers called clamshells, each one stuck with the Driscoll label. Before the clear plastic clamshells were green plastic baskets and before those, cardboard boxes.

Modern berry marketing goes back to 1912, when two berry growers in Pajaro Valley brainstormed a promotional idea. They were Richard Francis "Dick" Driscoll and Joseph Edward "Ed" Reiter, actually brothers-in-law (R. F. Driscoll was married to Ed's sister). Both were the only Banner strawberry producers in California and they wanted every commission man and buyer to know how special the Banner was. So they tied their berry crates with a blue paper ribbon that featured a lithographed image of a fancy red strawberry. The idea took off. This established the Banner as a recognized variety and fueled both Driscoll and Reiter's later efforts in marketing and advertising berries.

Driscoll Strawberry Associates, Inc., wasn't officially born until after World War II. At its core was the second generation of Driscolls and Reiters, as well as a few other Pajaro Valley growers. R. F. Driscoll's sons, Ned and Don, along with Kenneth Sheehy, Tom Porter, Swede Johnson, and Dr. Harold Thomas, started the research organization the Strawberry Institute of California, Inc., on November 16, 1944. In February 1953, Driscoll Strawberry Associates was officially incorporated with Ned, Swede Johnson, Joe Reiter, George Driscoll, and attorney Robert Dreher as the corporation's first directors. Driscoll was dealing with both the fresh and frozen strawberry market until about 1959. Experiencing the glut of berries in the frozen market in the late 1950s, the company sold its freezer and concentrated entirely on fresh strawberries. In 1966 the Strawberry Institute of California merged with Driscoll Strawberry Associates, thereby establishing Driscoll as a forerunner in patented strawberry varieties in the private sector.

So, in a sense, Driscoll Strawberry Associates (DSA) doesn't *grow* a single strawberry for sale, but it does all the research, distribution, and marketing for all the growers associated with DSA. It provides strict guidelines for when the plants need to be placed in the fields and gives advice on how to deal with pests and other problems. Looking back, I was well suited to work with Driscoll. I was always good at dealing with schedules and being on time. In fact, it was in following the company guidelines faithfully that I came up against my boss, Ned Driscoll, one day.

It was during my first year as general manager for E. F. Driscoll Farming Trust, one of DSA's main growers. The research department provided us with strict planting dates, and if we missed planting a certain variety at a certain date, whether because of rain or other circumstances, we had to plant another variety. I had heard that Ned and the foreman of his Salinas ranch had come to our Watsonville cooler to pull out some strawberry plants for planting. I went to the cooler to head him off.

"No," I told him. "Those plants are out of date." Those particular varieties were his favorite but, according to the company's planting guidelines, the planting date was past due.

"We've got to plant this one," he insisted.

But I kept telling him "no." Soon about fifteen to twenty men, including other growers and members of the research department, had gathered outside the cooler. One of them was my old friend John German, a former tomato inspector who was then the manager in charge of the plants in the cooler.

Ned told John to give the truck driver the plants, but he wouldn't go along with it because I hadn't said it was okay. But that didn't stop Ned. He demanded the keys to the cooler. Frustrated, John threw the keys on the table and said, "I quit."

That ended the conflict. John didn't quit and Ned didn't get his plants. I didn't like confrontations, but I couldn't walk away from that incident. If I walked away, there was no use in me having that job. A group of men were watching me and if I went along with Ned's demands, I would lose credibility. It would have been different if Ned had taken me aside and said, "It's only a week late, so let's take a chance." But instead, by making such a public display, he was making me and John German look bad.

After that incident, every time Ned would see me he would say, "I'm going to fire you." But he didn't say it in a serious way. All I did in response was smile.

CHAPTER

Removing Lemons

As my father was well aware, clearing the land of trees for crops is one of the most rigorous physical labors a farmer faces. When I acquired the Pidduck Ranch, an eighty-acre piece of land in Oxnard, in the 1970s I couldn't just set the existing lemon orchard on fire like my father had done with his willow trees in 1930. No, I had to do it the hard way: bulldoze the trees one by one.

Other than once being fertile ground for sugar beets, lima beans, and walnuts, Ventura County is probably most famous for its citrus groves, particularly lemons. Lemons were introduced in the area as early as 1883. Japanese immigrants, in fact, worked in the citrus groves for Rancho Sespe and the Limoneira Company, the latter of which still exists today. In 1927 the Ventura County Agricultural Commissioner reported 5,798 acres devoted to the production of lemons. By 1947, that number had grown to 19,570.

In the 1970s lemons were still the number one agricultural product in the Oxnard Plains. But strawberries were slowly starting to take their place. To get some help in clearing the Pidduck Ranch, I called on J. Miles Reiter, the older of Joe Reiter's two sons and a young leader of Driscoll Strawberry Associates.

Joe Reiter's father, Ed, as I mentioned before, was R. F. Driscoll's brother-in-law. Together the Reiters and Driscolls formed a powerful strawberry dynasty that has lasted three generations so far. Ned

Driscoll himself had both Driscoll and Reiter blood in him because his mother had been a Reiter.

Miles Reiter had lived on a Santa Clara berry farm amidst Japanese American sharecroppers until the age of six. "I don't know if I knew anybody that wasn't Japanese," he says, remembering picking raspberries for twenty-five cents a bucket for an Issei grower as a child. Later the Reiters moved off of the ranch to the suburban area of Los Altos, and Miles went on to earn a degree at Princeton University. After studying at an Ivy League college, he returned to farming in 1971 at the age of twenty-two, back in the fields and dirt hauling berries and surveying.

Back then we weren't using drip irrigation yet so it was important for the rows of strawberries to be perfectly level for the water to flow evenly through the entirety of the fields. Kay Mukai, one of DSA's pioneers, was one of the best natural surveyors I've ever seen. I was always surprised to see how fast he could straighten out a ranch to dead level, which means a straight line at 180 degrees. Kay was so fast that he wouldn't charge anything for surveying. Usually we'd have to hire an engineer from nearby Sunnyvale to do the same work.

Miles, on the other hand, learned surveying from Mike Miyakawa, who ran his father's farm operation. Miles's main teacher of other strawberry lessons was a Japanese sharecropper in his sixties named Susumu Mizuta. Miles recalls:

> *He taught me a lot about paying attention to plants, particularly with regard to irrigation. He was really a master irrigator. His attention to detail was incredible. No matter [what] level I had made as a surveyor, [if it was off,] he figured out ways to make it work. He would show me the bugs and how they multiply, and which ones to watch out for. He was just so committed and observant. He was strong and could be tough, but he was also a man of few words. We'd have these grower meetings, and*

he didn't say much, but boy, when he did, that could change everything. He was the first person I got to really watch shift a whole group of people with one sentence.

By the mid-1970s, Miles was a rising star in the Driscoll organization. His late father Joe had at one time started some pioneering farms in the Oxnard area with his manager, Mike Miyakawa, but nothing much had come out of it. Now Miles wanted to see if he could expand their strawberry holdings beyond Northern California to Oxnard. "Come on down," I told him. "Help me pull these lemons out and get this field ready to farm."

As it ended up, Miles didn't come down. But his younger brother Garland did. In a sense, each specialized in a different part of the state. With family farm operations, that's not a bad idea. That way each can have his space and own way of doing things yet still be part of the same larger organization.

After preparing the Pidduck Ranch, I found another thirty-acre piece of land in Oxnard for Driscoll strawberries and raspberries for the Reiters. It was called the Fuji Ranch and I heard the land was up for lease. The Fuji brothers were nurserymen in the Sawtelle area west of Los Angeles—largely a Japanese American community.

One day I went to talk to the Fuji brothers. We didn't know each other from Adam, so I was trying to establish a relationship. They were soft-spoken, kind, and ready to help us in any way. We talked for a while, and then Grandpa Fuji walked in. Wouldn't you know it, Grandpa Fuji knew my dad through some kind of celery plant connection. Needless to say, we got the ranch and even more—an option to buy.

CHAPTER

SHARECROPPING

VS.

SUBCONTRACTING

When I was a kid working on my father's garlic ranch, I heard the sharefarmers talk about my dad: "Look, he can put on a suit and go off while we work hard." Since I was always quiet, they didn't notice that I was there behind them, listening to what they were saying. My father didn't hear the comments and I didn't tell him. Sometimes I'd figured, "What the heck? Maybe he deserved it." But I knew deep down inside that he was probably going off to another meeting and that was part of his work, whether people realized it or not.

I never liked the word "sharecroppers" because of what people thought of it. In sharecropping, the main farmer provides the acreage, supplies, equipment, and even housing to families to work the land, and in exchange, the sharecropper hands over a percentage —a large amount in the worst cases—to the main land holder. When I became the general manager of E. F. Driscoll Farming Trust, I made it a priority to get rid of all Ned's sharecroppers. He had between forty and fifty on two big ranches. One of Ned's sons, Thomas or "Tommy," didn't like sharecropping either. Sharecropping once did help some people, including immigrants in the early 1920s and Japanese Americans returning from camps after World War II. At those times, people who didn't have any money or land could depend on a main farmer for strawberry plants and rent. All they needed to provide was labor.

While sharecropping served its purpose, it had a negative image. As the United Farm Workers were organizing sharecroppers into unions, we were faced with new labor laws and new definitions. The sharecropper, or sharefarmer, was supposed to incorporate his own culture into scheduling planting, irrigation, and harvesting. In essence, he was a farmer in his own right. But the minute the grower imposed his own scheduling and growing requirements, the sharefarmer would become his employee. Some farmers were hit with lawsuits because they required their sharefarmers to adopt certain growing methods but they weren't providing medical insurance and other benefits, as they would to regular employees.

I thought that it was best to get rid of the system altogether. I didn't do it cold; I gave the sharecroppers a couple of years warning. The attorney Robert Dreher, Bunny Driscoll, Tommy, and I visited the two ranches and we told them what was going to happen. Some were worried about where they were going to go, but we told them that if they wanted to continue sharecropping, we'd find them positions with other farmers. Others wanted to quit, and another group got into subcontracting for Driscoll associates. Subcontracting is different than sharecropping. When you're a subcontractor for a Driscoll associate, you finance your own farm and use Driscoll plants, but you have to pay the associate a royalty fee, actually a percentage of harvested crops, for the right to use the patented varieties. Not every grower can be an associate; a formal deal between the DSA organization and the grower helps the DSA choose its associates and maintain quality control. But once a grower is an associate, he can turn around and make subcontracting agreements with any farmer he trusts. Most of my subcontractors, in fact, went on to become official Driscoll growing associates. These included Howard Tao and the Sakai brothers in Watsonville and Mr. and Mrs. Jack Matsuoka in Santa Maria.

So I'm proud that I did away with sharecropping at E. F. Driscoll Farming Trust. It worked out for everyone concerned.

CHAPTER

PITCHING NETS

Every region has its own quirks and problems. It could be a certain kind of pest, regulatory law, or troublesome soil. In Oxnard and Orange County it was birds. Seagulls liked strawberries but we farmers didn't like half-eaten berries. Other types of birds thrived on strawberry seeds. Sometimes these birds could wipe out an entire matured crop, so we all had to depend on creative ways to keep them out without harming them at the same time.

Grace Sakioka, who was running a successful strawberry operation in Orange County with her husband, John, went to an Asian market to buy freshly beheaded chickens. She hung the carcasses around her fields and her bird problem—at least with the seagulls—was solved. Garland Reiter went for another tactic: a model airplane that chased the birds. I decided to go another way. I pitched nets all around my fields so birds couldn't fly into my rows of berries.

In terms of progressive culture, Orange County was ahead of Oxnard, and Oxnard was ahead of Northern California. In Orange County, I visited Paul Murata, George Murai, and Mits Nitta—real pioneers—to see what they were doing. As I mentioned before, I knew the Muratas through my dad's Kumamoto connection and Mits Nitta had been a senior at UC Davis when I had just entered the university in 1941.

But ultimately most of our improvements in cultural practices came from our individual districts. There were, of course, University of California strawberry specialists Victor Voth and Dr. Royce S. Bringhurst. They were good farm advisors. Voth studied horticulture, including pomology, at both UCLA and UC Davis. Pomology involves the study of growing apples and pears, the first fruits to be studied in a scientific way. Most people don't realize that strawberries—as well as apples, pears, and nectarines—are all members of the rose family. If you've ever grown roses, you know that roses need a cold winter to produce a good crop of flowers, and strawberries are the same way; the plants need to be chilled for a while before being planted. Usually plants from the cool mountains in Redding are taken down to Southern California and planted in November. This is called winter planting. The plants are dormant, so they usually don't bloom for a year. The fruit then shows up the following year.

But as I mentioned earlier, it's best to produce strawberries when no one else has them. So Voth had the idea of placing plants in cold storage and planting them in Southern California from July to September instead of November, in other words, summer planting. Then, that next spring, the plants begin to bloom right away. This way the fruit can be harvested earlier, from February through July. Now we would have fresh strawberries all the year around.

Voth also introduced sprinkler irrigation to Southern California. The usual method of furrow irrigation had its problems. First, you had to make sure the furrows were dug out at dead level so the water would flow evenly. Second, the standing water in furrows tended to get salty, a death sentence for strawberries. Eventually drip irrigation made its way to strawberry growing. I had heard of a young Sansei (a third-generation Japanese American), an Idaho native, Doug Mita, who had experience with drip irrigation.

More changes were on the horizon. As advised by Professor Stephen Wilhelm, we began fumigating the soil with methyl bromide during the preplanting stage, when we were discing the ground way before the arrival of any plants. In the 1970s we expanded the width of our strawberry beds. This was another change revolutionized by

Bringhurst and Voth. It used to be two rows on a forty-inch bed; to save space, we made room for four rows on a sixty-inch bed. We placed plastic over the beds and used a machine to make holes where the strawberries were to be planted. The plastic served two purposes: both drainage and warmth. The plastic shed excess rainwater off of the beds while also retaining heat during the winter plantings.

I listened to these university experts because even though I was a Driscoll grower, I was mostly handling university berries like the Lassen to fill in. Driscoll still didn't have a good variety for Oxnard, but that would eventually change.

CHAPTER

NEW COMMUNITY

When Sumi and I moved to Oxnard, we decided not to get too involved in local community groups. We had done our share in Gilroy and San Jose and we needed to take a break. The transportation in the suburb we lived in, Camarillo, was also poor at the time. Besides, I was trying to establish myself in new territory and Sumi was helping the kids adapt while also assisting me with bookkeeping and payroll.

But certainly there were groups to join. The Oxnard Buddhist Church, for instance, had a long history in the area. It was built on East Sixth Street in downtown Oxnard in 1929 through the fundraising efforts of thirty-five pioneering Japanese American families. Located along a road lined with eucalyptus trees leading to a sugar beet factory, the church was led by the Rev. Taiken Masunaga, who often likened the voice of Buddha to "a lion's roar," according to the church historian. The church was officially incorporated in 1936. Signing the incorporation papers were two Nisei, Tomio Yeto and Hanako Kato. Hanako had been born in a one-room hospital in Oxnard in 1915. During the forced removal of Japanese Americans on the West Coast, she had made sure that the church's property taxes were paid in full to ensure the building's return to the community. After the war, the temple housed returning Japanese American families for the following ten years.

Another community building was the Seventh Street Recreation Center. On October 27, 1936, the Japanese Association signed a twenty-year lease to occupy the building owned by the city of Oxnard. The *kendo* club and other Japanese American groups also used the facility before the war. In March 1942, the building was turned over to the Spanish-American Alliance and the Filipino community for a year. In 1943 the USO used the building to provide services for African American sailors. The city's Recreation Department then retained its operation until 1950, when a petition was filed in Superior Court by the Japanese American Citizens League of Ventura County and the city of Oxnard to have the Oxnard Recreation Department vacate the property according to the terms of the 1936 lease.

Although the Japanese Association was by then defunct, there was still the Oxnard Japanese American Citizens League, then led by Toby Otani. Otani argued that the city should honor its lease to the Japanese American community until 1956. "Our people spent a great deal of money improving the building," he said.

In March 1951, the "friendly" suit was decided. Superior Court Judge Anthony Brazil ruled in favor of the Japanese American Citizens League. For the next five and a half years, the Seventh Street Recreation Center was controlled by the Japanese American community.

Today there's no Seventh Street Recreation Center. On the same block stands Oxnard's modern Performing Arts Center. The Oxnard Buddhist Church, however, still exists at the former site of St. John's Lutheran Church at 250 South H Street. As in the early days, students meet at the church for Japanese language school. New activities such as *taiko* (drumming) and *kendo* (the martial art of Japanese swordsmanship) have been added. On some days, I'm sure, you can hear sounds emanating from the church like "a lion's roar."

CHAPTER

RIGHT VARIETY

The biggest challenge I had in moving down to Oxnard was the lack of a good Driscoll strawberry variety for the area. Plant breeder Harold Johnson was working hard in Watsonville to try to find something for the district. But it's a process that takes time.

Driscoll had plenty of high-chilling varieties that produced good fruit in Northern California, but they just didn't have the low-chilling varieties we needed. Oxnard has pretty mild weather so strawberries don't need to go through long periods of chilling before breaking dormancy. Johnson had a summer-planted variety called the D-4, but it didn't last too long because the fruit size was small. Next came the D-8, which was an improvement in terms of both production and quality, even though we sometimes got too much fruit at one time. These berries were also small, but it was all right because we used them for export to Japan, where small fruit was wanted.

A lot was riding on Harold Johnson. Dr. Thomas and Earl Goldsmith, Driscoll's first breeders, had retired in 1966, and it was up to Harold to keep the reputation going. He hired entomologist Richard Nelson and had additional help from Joe Espejo, a son of a Driscoll sharecropper, and Al Amorao. Even though Joe and Al had no formal education in plant horticulture, Harold could see both were hardworking and intelligent. His hunch paid off; both are credited in countless Driscoll patented varieties.

To create a variety, you first cross the seeds. For instance, if you want to produce a low-chilling type, you have to identify two kinds of low-chilling plants and breed them together. This sounds easy but this may be the start of a five- to seven-year process.

In Oxnard, in fact, Driscoll grows thirty thousand seedlings every year in their test plots. Out of that thirty thousand, the company saves about two hundred, which are called "selections." Out of the two hundred selections, five to seven are introduced to a stage called "time of planting." In most years, none of these selections become successful varieties.

During my time, Driscoll didn't have its own test fields. Instead, Driscoll growers took turns supporting test plots on their own land. I had about two acres reserved for this purpose. The Driscoll breeders would tell me when to plant which new selection and what to fertilize, the steps that make up time of planting. Once the plants began to grow, Al would come by on a daily basis. He was always with his camera and macro lens; he even took a photography class in Salinas to learn how to better capture detailed images of strawberries.

I was testing the E-26 variety. The E-26 is an everbearer, which is different than a summer-planted or winter-planted berry. You can plant an everbearer at practically any time and it will produce berries in about eighty days. The downside is that they can be vulnerable to certain diseases and environmental conditions, but they are good to close the gap between the summer- and winter-planted varieties, filling in between the end of Watsonville production and the beginning of Southern California winter planting.

I had an idea that if we could produce strawberries in November, we would have a good market. Because once it starts raining, no strawberries can be picked, and because Northern California can get its rainy season as early as September, it means there might not be any berries until Florida's season in December. As its turns out, the E-26 did fill the gap. Decades later, the E-26 is still being grown in Oxnard—lasting longer than me in this business.

CHAPTER

UNION AND PROTESTS

What no strawberry grower wants to talk about is the union, specifically the United Farm Workers (UFW). There are at least two sides to every story, and being that I was a farmer managing a lot of acreage, that's the side that I was coming from.

I always prided myself in being progressive in terms of handling both the culture and labor of my workers. I was one of the first to start unemployment and health insurance for my workers in the valley. I had also set up an employee bonus system, as well as seniority ratings for my longtime pickers. I logged in how many years each worker had been with me, and the following year, those with seniority were called back first. They also had a chance to get better positions, such as being part of the long-stem strawberry crew, which earned them more money. Some worked for me for fifteen years. I once had four sisters as part of my crew, two of them as timekeepers. I watched all of them grow up, get married, and eventually leave fieldwork.

I gave out the year-end bonuses based on how many hours they had clocked during the season. The bonuses could range from one hundred to two hundred and fifty dollars. I think I was the only one giving out those kinds of bonuses at that time.

So in the 1970s when labor leader Cesar Chavez was leading the United Farm Workers to organize migrant farm workers in first grape and later strawberry fields, I wasn't worried. Most of the organizing had taken place in Salinas and the Pajaro Valley, more than two hundred and fifty miles away. After a strike broke out in August 1970, members of Naturipe, Watsonville Berry Cooperative, and ten other growers and shippers started negotiations with the UFW. Many of the Japanese American farmers were from families that had once worked as either field workers or sharecroppers. Now the tables were turned.

It was only a matter of time before the union would target farms in Ventura County. The UFW wanted higher hourly wages as well as overtime pay, worker's compensation, and unemployment insurance. Some leaders were also concerned about the return of sharecropping after the *bracero* program had ended. Lawsuits were filed demanding a larger share of market proceeds and more control for the sharefarmers.

Since I was managing three hundred acres for E. F. Driscoll Farming Trust and we were part of Driscoll Strawberry Associates, it was no surprise that the union was going to come after our ranches. Driscoll was known as the largest shipper of fresh strawberries in the nation, and it's always easiest to go after the biggest target.

With our ranch leased from the Kita brothers located next to a highway, it was difficult for the union organizers to get access to our workers. I had fenced the field and the organizers picketed at our gate. They made a lot of noise, but I have to give them credit—they didn't say anything about me being Japanese. It would have been bad if it had become a racial battle. I hired security officers, but I also made sure I was around. I didn't want my workers to lose confidence in me.

Nothing came out of that first strike, but the second strike in May 1977 was a different story. At that time I had acquired a farm owned by John Maulhardt, and this ranch, unlike the Kita one, was open on all sides. In addition to me, other growers of strawberries and lemons along the coast in towns like Guadalupe, Arroyo Grande, and Lompoc were targeted. Groups like the Western Growers Association and the Ventura County Growers Association were mobilizing to help us farmers deal with labor organizing. The union wanted the farmers to increase our piece rates. A petition went around the ranch and enough workers signed it to hold a union election.

I figured that it was probably best if we just went through with the election. Some farmers hired labor consultants to fight the union, but I didn't want to get into that. On June 20, 1977, the Agricultural Labor Relations Board came by the Maulhardt and Kita ranches to conduct the election. A member of the United Farm Workers was also there to make sure everything was on the level.

Right before the vote, I had a chance to speak. I very seldom made speeches, but I spoke for close to half an hour. I reminded them about all I had done in terms of employee benefits and wages; their pay was already at the top of the scale. A UFW representative also made his case.

Then came the voting. Roger Smith, the field examiner supervising the election, tallied the votes. There were seventy-one votes for the United Farm Workers; ninety-nine for no union at all. E. F. Driscoll Farming Trust had beat the union. This was pretty unusual because most elections, close to 95 percent, resulted in a victory for the United Farm Workers. I eventually framed the yellow carbon copy of the tally sheet authorized by the Agricultural Labor Relations Board. It's completely faded now, but I still know it represents that my workers believed I had been fair to them.

CHAPTER

JAPAN MARKET

One of Japan's favorite desserts is a cake with strawberries on top. These cakes are cut so that a whole berry sits on top of each slice. That's why the Japanese buyers want small berries. I learned this and other aspects about the Japanese strawberry market when I went to Asia with Driscoll salespeople in the late seventies and eighties. The company wasn't looking to me to interpret, which was good because I couldn't communicate that well in Japanese. At times I was able to fill in the gaps. I gave the Driscoll executives my thoughts in English; sometimes they needed help understanding some of the cultural intricacies.

The Japanese like personal contact. You can mail letters and correspondence all you want, but if they don't get a chance to meet you, it's unlikely you will do much business together. Driscoll works mostly through brokerage houses that have big pipelines to supermarket chains, but we still had to go out and meet the principals of every company that handled our strawberries.

Before I went to Japan on these business trips, I had no idea there were so many layers of people in the berry market. By layers I mean the airline companies, transportation companies, and import license companies. Then there are the brokerage houses, supermarkets, and trucking companies. In all, there might be six or seven layers of businesses handling our commodity. In the United States, we usually hired employees to handle a lot of these tasks, but in Japan, different entities are part of the deal and all receive a certain percentage. It's no wonder it is so expensive to do business over there.

In Tokyo we went to a fruit auction house. The auction is held in a big warehouse where crates of fruit are rolled in on a conveyor belt. The buyers stand on bleachers holding paddles with numbers on them. The auctioneer announces the district where the fruit came from and then the bidding starts.

Certain agricultural districts in Japan have good reputations, so the bidding for that fruit starts out high. You won't believe how much people will spend on commodities like honeydew melons; I saw buyers purchase melons for seventy-five dollars each!

The strawberries came out in small cartons. You could immediately tell which regional associations produced the quality berries; some were nice and ripe; others were such a light pink color that they looked almost white. Most of the strawberries were sold through associations, not by individual growers. My father, for instance, had to buy into a co-op when he raised garlic and onions in southern Japan. When he raised strawberries, he had to go through another association.

In Japan, most of the strawberries are grown in greenhouses, and in summer and early fall, it gets too hot and the strawberries literally melt. If you figure that the fruit is 99 percent water and the skin is the only thing holding the flesh together, it makes sense that if you can't keep that skin solid, the strawberry will deteriorate pretty fast. It's during those hot months that the Japanese look to American strawberries. They like our berries because they are grown outside. In general, I think strawberries grown in the United States also taste better—less watery.

Even with superior berries, American growers may still face restrictive import laws. It's funny because when I was in farming, the Japanese government *required* American strawberry imports to be fumed with methyl bromide. They were afraid that we would otherwise bring spider mites into Japan. There are many other protective policies that affect agricultural imports. Japan has a certain way of doing things; it's not the American way. But you can't go into a foreign country and tell them how to conduct business.

In the end, it all comes down to quality. Since the Japanese have to deal with so many layers of vendors, they can't afford bad quality. And then there are those strawberry cakes. If you go to the railroad station, you'll see all kinds of people carrying boxes of these cakes as hostess gifts, also known as *omiyage*. When you go to someone's house, you always bring a gift. You never arrive empty-handed.

When I passed those cake stores in the train station, I couldn't help but feel something. Topping those cakes could be Driscoll berries from California. In this way, the strawberry played an important role in how the Japanese expressed their appreciation to friends and family.

CHAPTER  12

Entering the Circle

It was about 1977 when I began to feel tired. "I don't know if I have a future here," I told Bunny Driscoll.

I was getting a good salary with an annual bonus but I wasn't an equal among the other Driscoll principals. As fitting with the bylaws at the time, there were seven board members of Driscoll Strawberry Associates, Inc., and Tommy, Ned's son, was one of them. To be a formal Driscoll associate, you had to be a member of the board. I was still on the outside. Since I didn't have any shares of the private company, I really didn't have a voice.

But I was never intimidated by *Hakujin*, and I had added confidence because I had started a new deal for the company in Oxnard. I had found land and nailed down leases to good farmland. I did everything from the ground up.

I finally figured that I couldn't pay my dues to somebody my whole life. I had to progress. I knew that I had to get into the circle somehow. I saw that there were other growers who had been with the organization a lot longer than I had, yet they still weren't in the circle. If Driscoll Strawberry Associates wanted me around, they would have to offer me something.

I told Bunny Driscoll that the present arrangement wasn't working for me. So when one of the associates and founders of DSA, Swede Johnson, passed away, she approached his son-in-law, Clint

Miller, who was the president of the board at the time. "I don't want to take all of the stock," Clint told me. "I'll give you part of it." Later I learned that Clint's father-in-law, Swede, had a little bit of history with my dad. When my dad was sent to Bismarck, he and Mr. Rush made arrangements for Swede to take over one hundred acres of spinach that was ready to be harvested. Now his son-in-law was offering me my own shares of Driscoll stock.

The bylaws said that there could only be seven directors, but the board voted to change the bylaws to add an eighth director. So they made room for me, a virtual outsider. I wasn't related to anybody else. I was the first Japanese American to be on the board of Driscoll Strawberry Associates.

Ned Driscoll, called "Mr. Strawberry of California," passed away on November 3, 1981, after many years of bad health. He had done so much for the strawberry industry and so much for me personally. It was because of him that I got the chance to finally enter the circle.

CHAPTER  13

Manabi Farms and Long-Stem Strawberries

After Ned Driscoll passed away, E. F. Driscoll Farming Trust had to be reorganized. Part of Ned's leased acreage in Oxnard was transferred to me under a new incorporated entity, Manabi Farms. The board of directors told me, "Well, since you're taking Mr. Driscoll's portion in Oxnard, you can buy a portion of the stock."

We were all expecting this stock purchase. What the board didn't expect was my response.

"How much of that portion can you take?" they asked me.

"I want it all," I told them. They were dumbfounded. I had already gone to the bank and borrowed the money. I was prepared to buy all the shares that were available, which turned out to be 3.5 percent of the total held by Driscoll associates.

I incorporated my own company, Manabi Farms, to take over Ned Driscoll's Oxnard strawberry-growing operation. Since all the land was leased, everything was transferred over from E. F. Driscoll Farming Trust to Manabi Farms. A year later, the drip irrigation specialist I hired, Doug Mita, and I partnered up to create Casper Berry Farms, which lasted about a decade.

Manabi Farms, on the other hand, lasted until the early 1990s, built on one hundred and eighty acres that I had discovered and developed. I believe in progressing all the time. There's no sense in doing the same thing your whole life. Time is short. By taking on more acreage and more responsibility, I want to improve myself at each stage.

One of my specialties became long-stem strawberries. I started experimenting with them back in 1970, about the time I was getting to know another Oxnard transplant, Cecil Martinez. Born in Monterey Park in 1939, Cecil was about sixteen years younger than me. His father, an immigrant from the Philippines, had worked as a foreman on a vegetable farm in the Southern California community of Temple City. His mother was a Mexican American from Los Angeles.

When Cecil was ten, his family moved to Imperial Valley near the Mexico border. In the 1950s he attended Cal Poly, San Luis Obispo, a university known for agriculture and engineering. Instead of choosing to work with his father after graduation, Cecil came to Oxnard on his own in 1962.

I first met him out in the fields. I was driving in my pickup when I spotted him working on a vegetable farm. I stopped my truck, got out, and started talking to him. I learned a lot about people and farming just through small talk. While some people prefer to keep to themselves, I didn't mind sharing information. Later I would run into him at a cooler in town.

Back in the early seventies, there were only a couple of large coolers for fruits and vegetables in the Oxnard Plain area. Coolers were absolutely key to progressing in the strawberry business. There was no use in producing more berries if you have no place to store and cool them before shipping them out.

Coolers were also social centers for farmers. Usually, delivering strawberries is the last thing they do that day, so around two or three o'clock in the afternoon, people would gather to talk and play cards.

Cecil and I got along naturally. I can't quite put my finger on it, but I knew that we could work together. Then during a particular hard time in Cecil's life, he came to me. "Manabi, I don't know where to turn. I don't know who to go to. I need some help."

All along I had referred Cecil to different ranches, but then one day in the early 1980s, when I was launching Manabi Farms, I went to him again. "Come on," I told him. "I need you now. I have a plan to move forward."

I made Doug Mita in charge of one ranch and Cecil the manager of another. By this time the long-stem strawberries were going good. We call them long-stem because when we pick them we leave some of the stem beyond the green cap. But what distinguishes them more than the long stem is that these berries are the largest of the season. Since they are so unusually big, people like to serve them for special occasions. We pick them at the prime time, when the earliest strawberries are ripening, usually in spring in Oxnard. Strawberries grow large at first and then they go down in size throughout the season. As a result, we don't see stems in the summer in Oxnard. You have to get those in Watsonville, where the season has just started.

We usually pick long-stem strawberries on demand. I first got initiated into the long-stem business when a former Driscoll salesman, Bill Crowley, approached me about helping to prepare a half a dozen crates of special strawberries for President Ronald Reagan. They made such a splash that the strawberries were even mentioned in a newspaper article. Right after that, hotels began to request the long-stem strawberries. One of my key staff members, Steve Kawaguchi, was instrumental in pushing the stems. Still, in spite of the early promotion with Reagan, it didn't get going for quite a long time.

When Cecil joined us in the 1980s, we were regularly getting orders for stems from restaurants and event planners. We would send our pickers into the fields with a special box to go after the biggest berries. Cecil, however, was thinking of a new method of harvesting the long-stems. "There's a better way to do this, Manabi," he told me. "Okay, go for it," I told him.

Cecil organized the field workers into teams and told them that they would be paid according to how many boxes each team filled. So, in addition to an hourly wage, they were getting compensated by piecework. We got so productive that even other Driscoll growers began to consult with us on how we were handling the long-stem berries.

It was in the beginning of the 1990s when I approached my two foremen, Doug and Cecil. "I want to retire," I told them. I had already worked out a proposal on paper. I planned to split the equipment between the two of them and then over a period of some years, they could pay me back. The leased land would also be halved, but it would be up to them to finalize the financing. I had actually spoken to some bankers, so all Doug and Cecil had to do was talk to those I had referred them to. Like other men and women had done for me, I essentially opened doors for them.

I could have sold the operation for more money to a large corporation, but I wanted to create new opportunities for these younger men who had worked so hard for me. As it turns out, both of them left Driscoll to establish relationships with other berry distributors. I was disappointed, but it was their decision to make. I was busy preparing myself for new ventures, ventures that had little to do with strawberries.

CHAPTER

14

Longest Partnership, New Passions

As I mentioned earlier, five is my magic number. I spent about five years working for a San Jose produce company, five years working on my own in strawberries in Gilroy, and a little over five years as a foreman with E. F. Driscoll Farming Trust. Two cycles of fives, ten years, were spent operating Manabi Farms. The one thing that has definitely passed that five-year mark—many times over—is my marriage to Sumi.

It's hard for me to say how we have been able to make our marriage work for so many decades. Especially since my retirement, we spend a lot of time together. We like to travel and go see new things on the spur of the moment. I don't like to make hotel reservations, so sometimes that gets us into a bind when accommodations are tight. But it somehow works out.

My daughter, Marcia, says I'm unusual for a man in that I love shopping, even window-shopping. I'll take Sumi to any store she wants to go. For a while, I was collecting Japanese *netsuke* (figurines) and *yatate* (pen and ink sets) so we would frequent Japanese antique stores throughout California and even Asia. We enjoy movies and plays; Cecil Martinez used to always ask my opinion of the latest film released in theaters.

In terms of strawberry growing, Sumi has helped me in very specific ways, like overseeing payroll while I worked for Nod Driscoll during those early years in Oxnard. She was also out in the fields back in Gilroy, sometimes working in the packing shed. Since she was raised on a carnation farm in Mountain View, she was not adverse to hard work. She and her sisters were often called upon to debud the carnations and fasten the wire for the carnation plants. To this day, she likes keeping her hands busy and moving around. She learns by watching other people do things.

I guess we both are risk-takers. When I decided to go out on my own, there was no hesitation on her part. I could pursue any new opportunity I wanted. In the early 1980s a friend wanted me to invest in a new venture involving a cable station on the north shore of Reno. At that time, everyone thought cable TV was going to make it big. I did invest, and in about eighteen months, the station closed; it just couldn't compete with programming offered by other stations. Sometimes investments don't pan out.

In the early 1990s, I got involved in a Ferrari dealership in Los Gatos, not far from my hometown in Gilroy. Ever since I started driving at the age of fourteen, cars have been a passion of mine. I can remember certain years by the car I was driving at the time. One of my pride and joys was owning a Mercedes Gull Wing. So when this opportunity came up to invest in the Ferrari dealership, I jumped, head first. As a result of this connection, Sumi and I were able to visit the Ferrari factory in Italy. It was amazing, fancy all the way.

In addition to cars, I was also able to get involved in the motorcycle business. In the late 1960s, I had built a warehouse for McCormick Shilling on our former packing shed property in Gilroy. Then the company Nob Hill Groceries came in and built a 155,000-square-foot warehouse on the seven-acre property. At the end of December 1998, Nob Hill vacated the property, making it possible for Indian Motorcycles to move in.

Probably our biggest investment during our later years has been in the Japanese American National Museum. It all started, in fact, with the Japanese American World War II veterans, specifically Colonel Young Oak Kim, one of the leaders of the 442nd Regimental Combat Unit. He, with other veterans, was trying to establish a museum for veterans in Little Tokyo. Through Colonel Kim I met Nancy Araki, who was involved in helping to launch the fledgling museum. Before Sumi and I knew, we were pledging $1 million to the effort after Irene Hirano came on board as executive director of what then became the Japanese American National Museum. It was kind of crazy because we didn't know if this idea would work. In a sense, we were handing over money that would go toward a future house that could have replaced the one we had purchased in Camarillo back in 1965. (We, in fact, still live there today.)

One provision we insisted on at the time of the donation is that we remain anonymous. We, especially Sumi, don't like to put on airs. We figure that money and educational degrees don't make one person any better than another person. But then I had a funny experience. I was trying to help the museum raise additional seed money when one person said she doubted that anyone had really given $1 million. She thought it was just a come-on. I couldn't help but respond, "Well, I know that someone really did give that money, because I'm that person." Shortly thereafter, some administrators asked to go public with our donation because they had similar experiences with skeptics. So we did, and we have continued to give, mostly to the museum but also to the Go for Broke Education Fund and even some Japanese American film projects.

As Sumi says, it may be that Kumamoto streak in me. There was a time, when I was younger, when I used to get mad at my dad for giving too much to community organizations, and here I was, doing the same thing.

CHAPTER

FATHER AND SON

For about one month back in 1963, I stayed in Japan to be with my father. My mother had passed away by then and my father was spending a lot of time in the southern part of Japan trying to establish new ventures in garlic, onions, chili peppers, and strawberries. He lived in a part of Kumamoto that was completely flat. He was caught in a flood one season and from midnight to morning he walked in water to get out of the flood zone. He later said that's how he got his bad case of arthritis.

Between this arthritis and rheumatism, my dad was in a lot of pain. He'd come home to California to get prescription painkillers and then go back to Japan without doctors continuing to monitor his condition. In the winter of 1962 he became very sick and was sent to a university hospital in Kumamoto. That's about the time I went to visit him.

As the oldest son, I felt a certain amount of responsibility. Besides being worried about him, I also had to make sure his financial affairs were in order. He had made a lot of promises to Japanese investors. He had started to build a cooler for the strawberries and on the day he went to the hospital, a new car was delivered to his house. If special financing was needed, I was in the position to do it.

When I arrived to the hospital, my father was in a coma. The professors at the university hospital first didn't know what to do with him. They took him off all the painkillers except one. That brought him back and he was eventually transferred to a private hospital. One day before his sixty-third birthday on March 1, 1963, my father was finally released. I took him to his house in Kumamoto. As soon as I saw that he was back home and well cared for, I left for Oxnard. After some time passed, he returned to Gilroy, where he died on Christmas Eve.

My father and I probably had a typical relationship for an Issei man and his son. We weren't particularly close, in the sense that we didn't spend that much time together. My father never got mad at me. He never demanded too much. I did my own thing and never gave him directions. I'm probably the same way with my son, Mark. We do try to get together, but I don't tell him what to do. What impresses both Sumi and I is that our son has always been independent. Since he was a child, he has never asked us for anything.

My daughter, Marcia, lives with her husband, Steve Messinger, in wine country in Santa Rosa, California. Steve is like me in that he travels a lot for his employer, a multibillion-dollar wine company. Marcia studied dental hygiene at UC San Francisco and continues to work part-time in Santa Rosa.

My kids are private and low-key. They get embarrassed by some of our philanthropy because they see the Hirasaki name stamped on buildings. I've even donated money to institutions in their names. I have never really told them outright that I realize how much they've sacrificed for the family. They were uprooted from their home in Gilroy during a critical time in their lives. I saw the move to Oxnard as an adventure, but they probably had other thoughts in mind. Through it all, they never said anything. It's only now that I see that they probably had a hard time.

CHAPTER

Strawberry Diplomacy

Whenever a special event comes up, I make arrangements for Driscoll strawberries to be delivered. There's just something about a plump long-stem strawberry that makes people happy. Maybe that's why there are so many strawberry festivals across the country. The late Tsujio Kato started the California Strawberry Festival in Oxnard when he was mayor of the city in 1984. There are also festivals in Watsonville, Florida, Texas, Louisiana, and countless other locations. When some dignitary comes to town, other local growers and I are called in to bring premium California berries. In 1994 I even supplied Driscoll strawberries for a lunch held in honor of the Emperor and Empress of Japan. Perhaps, as my friend Nancy Araki says, it all goes back to the power of strawberry diplomacy.

In 1989, shortly after I stepped down from the board of directors of Driscoll Strawberry Associates, I became involved with a milestone event—a special dinner organized by the Japanese American National Museum called "America's Strawberry: Fruit of Our Labor," which commemorated Japanese American pioneers as well as university scientists Dr. Royce S. Bringhurst and Victor Voth. This dinner was notable because it joined the north and the south. Being a former Northern California grower who became a Southern California transplant, I was able to call on both sides to participate. Competing companies and cooperatives, different nationalities, the university and private researchers all came together to celebrate the Japanese American contributions to the strawberry industry. At one table was Miles Reiter, the president of Driscoll Strawberry Associates. At another sat the head of Naturipe Berry Growers. Driscoll researchers recognized the achievements of university scientists. We were all in the same room together, for the same purpose. Again, the power of strawberry diplomacy.

Many in attendance were third- or even fourth-generation strawberry growers but there were also some like me, who got into strawberries later in life. I'm sure that if everyone in that room had shared his or her experiences of growing strawberries, the stories wouldn't be all the same. For me, I've learned that it's important to know when to start new things and when to end them, too. You don't want to hang onto anything for too long. Everything has a cycle.

Having this point of view has helped through both good and bad times. I knew not to go overboard when we were making money and not to get too low when we weren't. Our forced move out of Gilroy during World War II didn't last forever, and when the time came for me to strike it out on my own, I left my hometown.

Through all of it, I've been lucky, I know it. On more than one occasion, my father's name has carried weight. But if I didn't go out there and try to open doors, these opportunities would have stayed hidden. Strawberries, for me, were my new deal. At the time I was a boy living on that garlic and vegetable ranch in Gilroy, I would have never guessed that my future lie in strawberries in Oxnard, hundreds of miles away. But that's the way it is with new deals. You can't predict them, but when they come along, you need enough instinct to grab hold of them and see where they take you.

CHRONOLOGY

The Life and Times of Manabi Hirasaki

1923   8 March. Manabi Hirasaki is born to Kiyoshi and Haruye Hirasaki in Gilroy, California, the oldest of eight children.

1930   Kiyoshi Hirasaki buys a 400-acre farm in Gilroy, California.

1931   Kiyoshi Hirasaki establishes a store, Hirasaki Seed, in San Jose's Japantown. The Hirasakis move to an apartment at the back of the seed store, where they live for a year before relocating to a home on the farm in Gilroy.

1940   The Golden Gate International Exposition is held on Treasure Island in San Francisco Bay. The Japan Pavilion is on display, and when the World's Fair ends, Kiyoshi Hirasaki, who has become the largest grower of garlic in Gilroy, purchases a portion of the pavilion.

1941   June. Manabi Hirasaki graduates from Gilroy High School.

1941   Summer. Kiyoshi Hirasaki transfers his farm operation into a trusteeship. Hardin Rush, a Gilroy-based farm implements dealer, and his wife, Elsie, are the trustees.

1941   Fall. Manabi Hirasaki enrolls at the University of California at Davis.

1941   September. Elements of the Japan Pavilion are rebuilt on Kiyoshi Hirasaki's Gilroy ranch by the Nishiura Brothers, carpenters based in San Jose.

1941   7 December. Japan bombs Pearl Harbor. Manabi Hirasaki decides to leave the university to return to his home in Gilroy.

1942   January. Kiyoshi Hirasaki is arrested by the Federal Bureau of Investigation and incarcerated at Sharp Park before being sent to an alien detention center in Bismarck, North Dakota.

1942   19 February. President Franklin D. Roosevelt signs Executive Order 9066, leading to the incarceration of all persons of Japanese ancestry in the Western region—both citizens and aliens—in concentration camps. For a period of three weeks, Japanese Americans are allowed to "voluntarily" move out of excluded military zones as long as they have either sponsors or an address to which they will relocate.

1942   March. Manabi Hirasaki and Hardin Rush travel to Grand Junction, Colorado, to find housing for the rest of the family during the course of World War II. Rush and his wife take care of the Gilroy farm and lease the land to neighboring farmers.

1942   4 April. The *bracero* program is instituted to bring laborers from Mexico to work on farms in the United States for a limited time. The program is in effect until the 1960s.

1942   Fall. Manabi Hirasaki attends classes at Mesa Junior College in Colorado.

1942   September. Kiyoshi Hirasaki is released from the Bismarck camp and is reunited with the rest of the Hirasaki family in Grand Junction.

1943   January. Manabi Hirasaki attempts to enlist in the Armed Forces. He goes to Denver for his physical but is turned away because he is of Japanese ancestry and is classified 4C, an enemy alien.

1943 1 February. President Roosevelt announces the formation of the 442nd Regimental Combat Team, an all-Nisei military regiment. Eligible Japanese Americans can now be called to active duty.

1943 April. Manabi Hirasaki is called to active duty by the United States Army. He reports to Camp Shelby, Mississippi, for basic training. He is assigned to the 522nd Field Artillery Battalion, C Battery, otherwise known as Charlie Battery.

1944 May. The 442nd Regimental Combat Team, including the 522nd Field Artillery Battalion, leaves for Europe.

1944 28 May. The 522nd Field Artillery Battalion arrives in Bari, Italy, and meets the balance of the 442nd Regimental Combat Team in Naples.

1945 8 March. The 522nd separates from the 442nd and heads toward Germany. Some members of the 522nd are present when Jewish death camp prisoners are liberated.

1945 7 May. Germany surrenders.

1945 Summer. The 522nd moves into a German town called Donauworth and the soldiers wait to be either reassigned to another unit or discharged.

1945 14 August. Japan surrenders.

1945 September. Manabi Hirasaki and 112 other 522nd soldiers from the mainland are transferred to the 53rd QM Base Depot near Nuremberg.

1946 January. Manabi Hirasaki is honorably discharged from the Army and returns to the United States. After spending two weeks with friends in Chicago, he returns to his hometown of Gilroy, where his family has reestablished their farming operation, which eventually expands to 1,200 acres.

1947 19 January. Manabi Hirasaki marries Sumi Iwata, the youngest daughter of Daemon and Jiu Iwata of Mountain View, California. The wedding ceremony is held at the San Jose Buddhist Church.

1948 February. The *Hokubei Mainichi*, a Japanese American bilingual daily newspaper based in San Francisco, is launched with Kiyoshi Hirasaki as one of the founding stockholders. This same year he becomes the newspaper's second president, a position he retains until his death in 1963.

1948 13 December. Mark Kiyoshi Hirasaki is born to Manabi and Sumi Hirasaki in Gilroy.

1950 Hirasaki Farms, Inc., ends its farming operation.

1950 Manabi Hirasaki begins work as a manager for Ghiselli Brothers, a produce shipper in San Jose, California.

1953 February. Driscoll Strawberry Associates (DSA) is officially incorporated.

1953 30 November. Marcia Haru Hirasaki is born to Manabi and Sumi Hirasaki in Gilroy.

1955 Spring. Manabi Hirasaki grows strawberries with a partner, Robert K. Byers, in Gilroy. He eventually expands into Watsonville.

1956 Manabi Hirasaki begins selling his university fresh berries to DSA.

1960 Spring. Haruye Hirasaki passes away in Japan. A memorial service is held at the San Jose Buddhist Church.

1963 24 December. Kiyoshi Hirasaki passes away at the Gilroy Hospital.

1965 Manabi Hirasaki relocates to Oxnard, California, to oversee the Southern California operations of Nod Driscoll. He is later appointed general manager of E. F. Driscoll Farming Trust, which includes strawberry farms in Salinas, Watsonville, Santa Maria, and Oxnard.

1978 Manabi Hirasaki becomes the first Japanese American to serve on the board of directors for DSA and is appointed to the California Strawberry Advisory Board. He also starts his own farming operation, Manabi Farms, Inc., which continues in Oxnard until 1994.

1987 Manabi Hirasaki joins the board of trustees for the Japanese American National Museum. He later assumes various other leadership positions within the organization.

1989 Manabi Hirasaki is a partner in Casper Berry Farms in Oxnard. The partnership ends in 1994.

1990 4 November. The Japanese American National Museum holds the "America's Strawberry: Fruit of Our Labor" dinner.

1991 Manabi Hirasaki receives a medal from the Japan Agricultural Society (Nokai) for his contributions in agriculture.

1992 The Manabi and Sumi Hirasaki Family Theater Gallery is named at the Japanese American National Museum.

1999   January. The Manabi and Sumi Hirasaki Family Garden is named at the Japanese American National Museum's new Pavilion.

1999   December. The Manabi and Sumi Hirasaki National Resource Center is also named at the Pavilion.

2003   December. *A Taste for Strawberries: The Independent Journey of Nisei Farmer Manabi Hirasaki* is released.

Acknowledgments

We would like to thank all those individuals who kindly agreed to meet with and grant interviews to the co-writer, Naomi Hirahara, and researcher, Teru Kanazawa Sheehan. First of all, acknowledgments should go to past and present members of the Driscoll organization in the Watsonville area: Miles Reiter, Thomas Driscoll, Harold Johnson, Richard Uyematsu, Ken Morena, and John German. Also, within Driscoll's Oxnard operation, Amado "Al" Amorao provided information on the regional strawberry varieties in Ventura County, as well as beautiful color slides and prints of photographs of strawberries he had taken in the 1970s and '80s. Miles Reiter, Harold Johnson, and Al Amorao were particularly helpful in reviewing the manuscript and offering comments.

Those with memories about Kiyoshi Hirasaki and his associates included Dennis "Denny" Donovan, the late Kiyoshi Nishiura, and Matsue Kami. Giving personal recollections were Sumi Hirasaki, Michiko (Hirasaki) Sakamoto, and Marcia (Hirasaki) Messinger. The co-writer and researcher were also taken on a tour of the Japanese Pavilion on the former Hirasaki ranch by Mineko (Hirasaki) Sakai, who also generously provided articles on Kiyoshi and Haruye Hirasaki. Painting a picture of the Gilroy and Watsonville area included Masaru "Moose" Kunimura, Hiromi Nagareda, Charles and Nancy Iwami, Mary Okamoto, and the Rev. Sumio Koga.

The Japanese American National Museum's archive of strawberry interviews, conducted by Nancy Araki and Vernon Takeshita in the late 1980s, were extremely helpful in giving a full context of strawberry growing throughout the West Coast. Additional interviews with

Cecil Martinez, Grace Miye (Nakayama) Sakioka, Shoichi and Helen Kobara, and Yonejiro Ito were also illuminating. Background of the founding of the *Hokubei Mainichi* was told by Michi Onuma, while the late Shiro "Juggie" Takeshita and George Ishihara detailed their experience with the 522nd Field Artillery Battalion.

Also helping with resource material were Carol Champion of Special Collections at the University of California at Santa Cruz, Jane Borg of the Pajaro Valley Historical Association, and the Gilroy Historical Museum.

This published work would not have been possible without the Japanese American National Museum's commitment to tell the story of Japanese Americans. We acknowledge President and Chief Executive Officer Irene Hirano as well as Akemi Kikumura-Yano and Karin Higa. The book was ably coordinated by Carla Tengan and Susan Chen, with initial interviews collected under the supervision of Darcie Iki. John Esaki, Carol Komatsuka, Maria Kwong, Florence Ochi, Cris Paschild, and the curatorial and programs staff provided essential guidance and assistance.

Comments from Arthur Hansen, Lane Hirabayashi, Brian Niiya, Jim Hirabayashi, and Daniel Lee also helped to further refine the text. Patricia Wakida of Heyday Books generously shared some of her publishing expertise. Lisa K. Manwill gracefully and thoroughly edited the book. For the final transformation from text and images into a tangible product, we credit the expertise of designer Qris Yamashita, who was assisted by Phillip Komai and Russell Oshita, with new photography by Norman Sugimoto.

And of course, a special thanks again go to Sumi Hirasaki, Mark Hirasaki, and Marcia and Steve Messinger. Without you all, life wouldn't be as sweet.

Sources

The photographs reproduced in this volume are from the private collection of Manabi and Sumi Hirasaki, unless otherwise indicated.

Oral History Interviews

All of the following oral history interviews, conducted by the author and/or researcher, Teru Kanazawa Sheehan, are part of the permanent collection of the Japanese American National Museum:

Amado "Al" Amorao (30 July 2002, Oxnard, Calif.)

Dennis "Denny" Donovan (3 December 1998, Salinas, Calif.)

John German (5 November 1999, Watsonville, Calif.)

Manabi Hirasaki (15 September 1998, Camarillo, Calif.), (23 October 1998, Camarillo, Calif.), (26 October 1998, Camarillo, Calif.), (2 November 1998, Camarillo, Calif.), (17 November 1998, Los Angeles), (11 February 1999, Los Angeles), (26 February 1999, Los Angeles), (15 March 1999, Los Angeles)

Sumi (Iwata) Hirasaki (7 February 2002, Los Angeles)

George Ishihara (28 November 1998, Los Angeles)

Yonejiro Ito (5 February 1999, Oxnard, Calif.)

Charles and Nancy Iwami (10 November 1999, Watsonville, Calif.)

Harold Johnson (5 November 1999, Watsonville, Calif.), (7 November 1999, Watsonville, Calif.)

Matsue Kami (6 November 1998, Los Angeles)

Shoichi and Helen Kobara (9 November 1999, Watsonville, Calif.)

Masaru "Moose" Kunimura (3 October 1998, Gilroy, Calif.)

Cecil Martinez (28 October 1999, Oxnard, Calif.)

Marcia (Hirasaki) Messinger (5 December 1998, San Francisco)

Ken Morena (3 December 1998, Watsonville, Calif.)

Hiromi Nagareda (4 December 1998, Gilroy, Calif.)

Kiyoshi Nishiura (4 December 1998, San Jose, Calif.)

Mary Okamoto (4 November 1999, Watsonville, Calif.)

Michi Onuma (6 November 1999, San Francisco)

Miles Reiter (3 December 1998, Watsonville, Calif.)

Michiko (Hirasaki) Sakamoto (6 November 1999, Berkeley, Calif.)

Grace Miye (Nakayama) Sakioka (14 October 1999, Orange County, Calif.)

Shiro "Juggie" Takeshita (5 December 1998, San Leandro, Calif.)

Richard Uyematsu (7 November 1999, Watsonville, Calif.)

In 1988 and 1989 the Japanese American National Museum collected oral histories of men and women involved in the strawberry industry. The interviews are also part of the National Museum's permanent collection. Collecting the life histories were Nancy Araki and Vernon Takeshita, unless indicated otherwise:

Herb Baum (27 June 1989, Watsonville, Calif.)

Ed Domoto (31 December 1988, Fresno, Calif.)

Timothy Driscoll (26 June 1989, Salinas, Calif.)

Larry B. Galper (26 June 1989) (29 June 1989, Watsonville, Calif.)

Tak Higuchi (5 January 1989, San Jose, Calif.)

Manabi Hirasaki (26 March 1989, Camarillo, Calif.)

Tomio Ito (29 June 1989, Orange County, Calif.)

Ken Kitasako (25 February 1989, Nancy Araki and Fred Hoshiyama

George Miyake Jr. (13 May 1989)

Tom and Harumi Murakami (25 June 1989, Watsonville, Calif.)

Paul Murata (23 June 1989, Vernon Takeshita, Calif.)

Masaru Ronald "Buzz" Noda (24 June 1989, Watsonville, Calif.)

Miles Reiter (27 June 1989, Watsonville, Calif.)

David Riggs (28 June 1989, Watsonville, Calif.)

Terrance Sheehy, Robert Sheehy, and Patrick Sheehy with Kiyoshi Nishimori (27 May 1989, Santa Maria, Calif.)

Hiroshi and Chieko Shikuma (27 June 1989, Watsonville, Calif.)

Esau Shimizu (5 January 1989, San Jose, Calif.)

Kuni Shinta (26 June 1989, Watsonville, Calif.)

Ted Takahashi (31 May 1989)

Fumi Uyeshima (27 May 1989, Santa Maria, Calif.)

Victor Voth (23 June 1989, Vernon Takeshita, Irvine, Calif.)

Cha Young (31 May 1989, Fresno, Calif.)

PART ONE: FALL

Chapter One: Burning Creek

10 *Japanese first started farming:* Timothy J. Lukes and Gary Y. Okihiro, *Japanese Legacy: Farming and Community Life in California's Santa Clara Valley* (*Local History Studies* 31) (Cupertino, Calif.: California History Center, 1985).

10 *The daughter, Matsue Kami, explained:* Interview with Matsue Kami (6 November 1998, Naomi Hirahara and Teru Kanazawa Sheehan, Los Angeles).

Chapter Two: Growing Seeds

13 *Yoshio Nagareda first came to California:* Interview with Hiromi Nagareda (4 December 1998, Naomi Hirahara and Teru Kanazawa Sheehan, Gilroy, Calif.).

13 *The owner, Lin Walker Wheeler: Gilroy's First Century of Incorporation, 1870-1970: A History of the City* (Gilroy, Calif.: City of Gilroy and Gilroy Historical Society, 1970): 20.

14 *My friend Hiromi once explained:* Interview with Hiromi Nagareda.

14 *His father, Kazuto, worked for Wheeler's competitor:* Interview with Masaru "Moose" Kunimura (3 October 1998, Naomi Hirahara and Teru Kanazawa Sheehan, Gilroy, Calif.).

16 *That market is still there:* "Dobashi Market Is Still Going Strong," *Hokubei Mainichi*, 1 January 1990; Billie Lee, "San Jose Japantown Businesses: It's a Family Affair," *Nichi Bei Times*, 1 January 1998: 1.

Further Reading

Phyllis Filiberti Butler, *Old Santa Clara Valley: A Guide to Historic Buildings from Palo Alto to Gilroy* (San Carlos, Calif.: Wide World Publishing/Tetra, 1996).

Edgar J. Clissold, *The Seed Industry* (No. 3 in the Series of American Industries) (New York: Bellman Publishing Company, Inc., 1946).

A. J. Wells, "Seed Farms in California," *National Geographic Magazine* (May 1912).

CHAPTER THREE: ONE OF A KIND

17 *My Number Three sister, Michiko:* Interview with Michiko (Hirasaki) Sakamoto (6 November 1999, Naomi Hirahara and Teru Kanazawa Sheehan, Berkeley, Calif.).

18 *When Mrs. Kami's first husband:* Interview with Matsue Kami (6 November 1998, Naomi Hirahara and Teru Kanazawa Sheehan, Los Angeles).

CHAPTER FOUR:
CIGARS, SMOKE, AND THE *ICHIBAN* MEN

19 *According to my friend Moose:* Interview with Masaru "Moose" Kunimura.

19 *According to community historian Patricia Baldwin Escamilla:* Patricia Baldwin Escamilla, *Gilroy, California: A Short History* (pamphlet) (Gilroy, Calif.: Gilroy Historical Museum): 20.

19 *Moose remembers relations between the Chinese and the Japanese:* Interview with Masaru "Moose" Kunimura.

CHAPTER FIVE: LESSONS

FURTHER READING

Eiichiro Azuma, "Interethnic Conflict under Racial Subordination: Japanese Immigrants and Their Asian Neighbors in Walnut Grove, California, 1908–1941," *Amerasia Journal* 20, no. 2, 1994: 27–56.

Emory S. Bogardus, "Anti-Filipino Race Riots," *Letters in Exile* (Los Angeles: UCLA Asian American Studies Center and Regents of University of California, 1976): 51–62.

Howard A. DeWitt, "The Filipino Labor Union: The Salinas Lettuce Strike of 1934," *Amerasia Journal* 5, no. 2 (1978): 1–21.

Howard A. DeWitt, "The Watsonville Anti-Filipino Riot of 1930: A Case Study of the Great Depression and Ethnic Conflict in California," *Southern California Quarterly* 61, no. 3 (Fall 1979): 291–302.

Chapter Seven: Driving

28 *The senior Kishi, Kichimatsu, started:* Gwendolyn Wingate, "The Kishi Colony," in *The Folklore of Texan Cultures*, Francis Edward Abernethy, ed. (Austin, Texas: The Encino Press, 1974).

29 *According to historian Patricia Baldwin Escamilla:* Patricia Baldwin Escamilla, *Gilroy, California: A Short History* (pamphlet) (Gilroy, Calif.: Gilroy Historical Museum): 19.

29 *Mr. Sakata bought the springs:* Jeanne Patterson, "A Historical Biography of Harry Kyusaburo Sakata, 1885–1971," (unpublished, class paper, UCLA, 12 June 1977).

Chapter Eight: Garlic King

30 *Yoshio Nagareda was one of the early garlic planters:* Interview with Hiromi Nagareda.

30 *Back in the 1890s:* Patricia Baldwin Escamilla, *Gilroy, California: A Short History* (pamphlet) (Gilroy, Calif.: Gilroy Historical Museum): 21.

30 *By 1938, 50 percent of the garlic:* Masakazu Iwata, *Planted in Good Soil: A History of the Issei in United States Agriculture* (New York: Peter Lang, 1992): 323.

32 *One of our subcontractors were the Nakayamas:* Interview with Grace Miye (Nakayama) Sakioka (14 October 1999, Naomi Hirahara, Orange County, Calif.).

33 *Grace Sakioka remembers:* Ibid.

33 *According to* Smithsonian *magazine:* Richard Wolkomir, "Without Garlic, Life Would Be Just Plain Tasteless," *Smithsonian* 26, no. 9 (December 1995).

33 *As it states in a 1997 publication by the Gilroy Historical Museum:* Patricia Baldwin Escamilla, *Gilroy, California: A Short History* (pamphlet) (Gilroy, Calif.: Gilroy Historical Museum): 22.

FURTHER READING

Roy D. McCallum, "Growing and Handling Garlic in California," *Agricultural Extension Service* 84 (February 1934): 3–16.

"Sixteen Who Left Their Mark on Gilroy: Hall of Famers Honored Tonight," *Gilroy Dispatch*, 11 September 1987: C1, C6.

CHAPTER NINE: JAPAN PAVILION

34   *What we wanted to see was the Japan Pavilion:* Naomi Hirahara, "Master Artisans of San Jose: The Nishiura Brothers," *Japanese American National Museum Magazine*, Winter 2000.

34   *The pavilion, literally a large castle-like building with a three-level pagoda:* "Welcome to the Japan Pavilion" (pamphlet) (San Francisco: Golden Gate International Exposition, 1940).

35   *Approximately 4.5 million visitors:* Ibid.

FURTHER READING

Clay Lancaster, *The Japanese Influence in America* (New York: Walton H. Rawls, 1963): 179–181.

Official Guide Book, Golden Gate International Exposition on San Francisco Bay, 1940.

Dale Rodebaugh, "Historic Step for Japanese Home," *San Jose Mercury News*, 30 March 1987: 1B, 6B.

CHAPTER TEN: MR. RUSH

FURTHER READING

"Many Apply for Japanese Cherry Trees," *Register-Pajaronian*, 3 May 1939.

"More Japanese Cherry Trees to Be Given in 1940," *Register-Pajaronian*, 3 June 1939.

PART TWO: WINTER

CHAPTER ONE: ON THE MOVE

53 *The government designated this area: Report of the Commission on Wartime Relocation and Internment of Civilians, Personal Justice Denied* (Washington, D.C.: U.S. Government Printing Office, 1982): 100-104.

FURTHER READING

Leonard J. Arrington, "Utah's Ambiguous Reception: The Relocated Japanese Americans," in *Japanese Americans: From Relocation to Redress*, Roger Daniels, Sandra D. Taylor, and Harry H. L. Kitano, eds. (Salt Lake City, Utah: University of Utah Press, 1986): 92–97.

Roger Daniels, "The Forced Migrations of West Coast Japanese Americans, 1942–1946: A Quantitative Note," in *Japanese Americans: From Relocation to Redress*: 72–74.

Arthur Hansen, "James Matsumoto Omura: An Interview," *Amerasia Journal* 13, no. 2 (Fall/Winter 1986): 110–113.

Randall K. Sakamoto, "Forced Volunteers: An Oral History of Japanese American Voluntary Evacuation" (unpublished, senior thesis, Claremont McKenna College, 1997).

Page Smith, *Democracy on Trial: The Japanese American Evacuation and Relocation in World War II* (New York: Simon and Schuster, 1995).

Frank J. Taylor, "The People Nobody Wants," *The Saturday Evening Post*, 9 May 1942.

CHAPTER TWO: GRAND JUNCTION

57 *As it turned out, less than two thousand: Report of the Commission on Wartime Relocation and Internment of Civilians, Personal Justice Denied* (Washington, D.C.: U.S. Government Printing Office, 1982): 103.

Further Reading

Rocky Nippon, March–September 1942.

Paul Shinoda, "Paul Shinoda," in *And Justice for All: An Oral History of the Japanese American Detention Camps*, John Tateishi, ed. (New York: Random House, 1984).

Chapter Three: Bismarck, North Dakota

59 *Italian and German seamen who were stranded*: Stephen Fox, *The Unknown Internment: An Oral History of the Relocation of Italian Americans During World War II* (Twayne Oral History Series, 4) (Boston: Twayne Publications, 1990).

Further Reading

John Christgau, *Enemies: World War II Alien Internment* (Ames, Iowa: Iowa State University Press, 1985).

Lawrence DiStasi, *Una Storia Segreta: The Secret History of Italian American Evacuation and Internment during World II* (Berkeley: Heyday Books, 2001).

Carol Van Valkenburg, *An Alien Place: The Fort Missoula, Montana, Detention Camp 1941–1944* (Missoula, Montana: Pictorial Histories Publishing Company, Inc., 1995).

Chapter Four: Enlisting

61 *The Nagaredas were in Poston I*: Interview with Hiromi Nagareda.

62 *My friend Moose Kunimura*: Interview with Masaru "Moose" Kunimura.

CHAPTER FIVE: CAMP SHELBY AND
THE 522ND FIELD ARTILLERY BATTALION

65 *Within Charlie Battery were approximately ten different sections:* Historical Album Committee of the 522 Field Artillery Battalion of the 442 Regimental Combat Team, *Fire for Effect: A Unit History of the 522 Field Artillery Battalion* (Honolulu, Hawaiʻi: Fisher Printing Co., 1998): 9.

66 *Members of the 442nd, in fact, went to Rohwer:* Ibid., 9.

FURTHER READING AND VIEWING

Thelma Chang, *We Can Never Forget: Men of the 100th/442nd* (Honolulu, Hawaiʻi: SIGI Productions, Inc., 1991).

Charlie Battery, *522 FABN, 1943–1945: A Legend* (booklet published in commemoration of the Charlie Battery Mini-Reunion) (San Francisco, Calif.: October 1991).

Lyn Crost, *Honor by Fire: Japanese Americans at War in Europe and the Pacific* (Novato, Calif.: Presidio Press, 1994).

Interview with Manabi Hirasaki (17 April 1999, Christine Sato, Camarillo, Calif.) Hanashi Oral History Program, Go for Broke Foundation.

Interview with George Ishihara (28 November 1998, Naomi Hirahara and Teru Kanazawa Sheehan, Los Angeles, Calif.).

CHAPTER SIX: MANEUVERS

68 *The wire and radio section:* Charlie Battery, *522 FABN, 1943–1945: A Legend* (booklet published in commemoration of the Charlie Battery Mini-Reunion) (San Francisco, Calif.: October 1991).

68 *Connected to our section was someone who grew up near Gilroy, Shiro "Juggie" Takeshita*: Interview with Shiro "Juggie" Takeshita (5 December 1998, Naomi Hirahara and Teru Kanazawa Sheehan, San Leandro, Calif.).

CHAPTER SEVEN: NAPLES

71 *Juggie was there and remembered:* Interview with Shiro "Juggie Takeshita.

72 *It took us about twenty-six days:* Ted Tsukiyama, "The 522nd Field Artillery Battalion," in *Fire for Effect: A Unit History of the 522 Field Artillery Battalion* by the Historical Album Committee of the 522 Field Artillery Battalion of the 442 Regimental Combat Team, (Honolulu, Hawai'i: Fisher Printing Co., 1998): 31.

CHAPTER EIGHT:
MINEFIELDS AND FORWARD OBSERVERS

73 *The forward observer unit:* George Oiye, "My Life with Charlie Battery," in *Fire for Effect: A Unit History of the 522 Field Artillery Battalion* by the Historical Album Committee of the 522 Field Artillery Battalion of the 442 Regimental Combat Team, (Honolulu, Hawai'i: Fisher Printing Co., 1998): 168.

75 *Only two men in our battalion were killed:* Ibid., 55.

CHAPTER NINE: MP IN MONTE CARLO

76 *The 442nd was pretty mangled:* Thelma Chang, *We Can Never Forget: Men of the 100th/442nd* (Honolulu, Hawai'i: SIGI Productions, Inc., 1991): 158–59.

CHAPTER TEN: DACHAU

78 *Once we hopped over the Rhine River:* Lyn Crost, *Honor by Fire: Japanese Americans at War in Europe and the Pacific* (Novato, Calif.: Presidio Press, 1994): 240; Thelma Chang, *We Can Never Forget: Men of the 100th/442nd* (Honolulu, Hawai'i: SIGI Productions, Inc., 1991): 162.

78 *They were the ones who came across the Dachau death camp:* Chester Tanaka, *Go for Broke: A Pictorial History of the Japanese American 100th Infantry Battalion and the 442nd Regimental Combat Team* (Richard, Calif.: Go for Broke, Inc., 1982): 117.

FURTHER READING

Michael Coit, "Japanese-American Veterans Recall Liberating Camp at Dachau," *Los Angeles Daily News*, 30 April 1995.

Chapter Eleven: Coming Home

80 *In September 1945:* "113 Men Transferred to QM Outfit: Mainland Boys Part with Battalion," *High Angle*, 22 September 1945.

Chapter Twelve: Hirasaki Farms, Inc.

82 *The new president of the seed company:* Correspondence dated 8 September 1944, from J. B. Scherrer, Pieters-Wheeler Seed Company, Gilroy, California to Hiromi Nagareda, Poston, Arizona.

83 *Hiromi eventually left:* Interview with Hiromi Nagareda.

83 *The Uyematsus, a family who had farmed strawberries:* Interview with Richard Uyematsu (7 November 1999, Naomi Hirahara, Watsonville, Calif.).

84 *In fact, according to Kazuko Nakane's research:* Kazuko Nakane, *Nothing Left in My Hands: An Early Japanese American Community in California's Pajaro Valley* (Seattle, Wash.: Young Pine Press, 1985).

Chapter Thirteen: Romance and Pickles

85 *Their father, Daemon:* Interview with Sumi (Iwata) Hirasaki (7 February 2002, Naomi Hirahara, Los Angeles).

Chapter Fourteen: Founding the *Hokubei Mainichi*

88 *One was the* Nichi Bei Shimbun: Steve Fugita, "Kyutaro Abiko (1865–1936)," *Distinguished Asian Americans*, Hyung-chan Kim, ed. (Westport, Conn.: Greenwood Press): 1–4; "Kyutaro Abiko: A Man with a Dream for the Japanese in America," in *Japanese American Journey: The Story of a People*, Florence M. Hongo and Miyo Burton, eds. (San Mateo, Calif.: Japanese American Curriculum Project, Inc., 1985): 107–111.

88 *Another daily newspaper was* Shin Sekai Asahi: "Growing Up with the Community: A Brief History of the *Hokubei*," *Hokubei Mainichi* (30 September 1998): 1–2.

88 *Even though our family had some personal ties:* Interview with Matsue Kami (6 November 1998).

89 *In 1939 a printer named Shigeki Oka:* Interview with Michi Onuma (6 November 1999, Naomi Hirahara and Teru Kanazawa Sheehan, San Francisco).

89 *That same year, Yasuo Abiko:* Nichi Bei Times Web site, accessed 15 May 2002.

89 *According to newspaper historians:* "Growing Up with the Community: A Brief History of the *Hokubei*," *Hokubei Mainichi* (30 September 1998): 2.

FURTHER READING

Daniel K. Yoo, "Reading All About It: Race, Generation, and the Japanese American Ethnic Press, 1925–41," *Amerasia Journal* 19, no. 1 (1993): 69–92.

CHAPTER FIFTEEN: OUR FARMS

91 *We had grown from six hundred acres:* "Celery: Hirasaki Farms, Gilroy, California" (promotional booklet), circa 1950.

91 *We already had come up with a name:* "Kiyoshi Hirasaki: The Garlic King," in *Japanese American Journey: The Story of a People*, Florence M. Hongo and Miyo Burton, eds. (San Mateo, Calif.: Japanese American Curriculum Project, Inc., 1985): 107–111.

91 *The fields would be disced at least several times*: "Celery: Hirasaki Farms, Gilroy, California" (promotional booklet), circa 1950.

92 *Later, a Chicago transplant named Denny Donovan Sr.:* Interview with Dennis "Denny" Donovan (3 December 1998, Naomi Hirahara and Teru Kanazawa Sheehan, Salinas, Calif.).

93 *The spiral-bound booklet featured photographs:* "Celery: Hirasaki Farms, Gilroy, California," (promotional booklet), circa 1950.

PART THREE: SPRING

CHAPTER TWO: STRAWBERRY DEAL

104 *Before the war, 90 percent of the strawberry crop:* Miriam J. Wells, *Strawberry Fields: Politics, Class and Work in California Agriculture* (Ithaca, New York: Cornell University Press, 1996): 112.

104 *He grew some acreage in Salinas:* Stephen Wilhelm and James E. Sagen, *A History of the Strawberry: From Ancient Gardens to Modern Markets* (University of California Division of Agricultural Sciences, 1974): 197.

105 *A year later, according to George M. Darrow's classic book,* The Strawberry: George M. Darrow, *The Strawberry: History, Breeding and Physiology* (New York: Holt, Rinehart and Winston, 1966): 23.

105 *A lot of things contributed:* Miriam J. Wells, *Strawberry Fields: Politics, Class and Work in California Agriculture* (Ithaca, New York: Cornell University Press, 1996): 112.

CHAPTER THREE: THE CRASH OF 1957

106 *Sho Kobara, who grows both apples and strawberries:* Interview with Shoichi Kobara (9 November 1999, Naomi Hirahara, Watsonville, Calif.).

106 *By the 1950s:* Miriam J. Wells, *Strawberry Fields: Politics, Class and Work in California Agriculture* (Ithaca, New York: Cornell University Press, 1996): 112.

106 *Naturipe's roots go back all the way to 1917:* Stephen Wilhelm and James E. Sagen, *A History of the Strawberry: From Ancient Gardens to Modern Markets* (University of California Division of Agricultural Sciences, 1974): 208; Lane Hirabayashi, *The Delectable Berry: Japanese American Contributions to the Development of the Strawberry Industry on the West Coast* (pamphlet) (Los Angeles: Japanese American National Museum, 1989).

107 *Smaller growers joined the Watsonville Berry Cooperative:* Interview with Masaru Ronald "Buzz" Noda (24 June 1989, Nancy Araki and Vernon Takeshita, Watsonville, Calif.).

Chapter Four: The Banner, Shasta, and Lassen

108 *Prior to the 1920s:* George M. Darrow, *The Strawberry: History, Breeding and Physiology* (New York: Holt, Rinehart and Winston, 1966): 211.

108 *The Banner strawberry variety:* Stephen Wilhelm and James E. Sagen, *A History of the Strawberry: From Ancient Gardens to Modern Markets* (University of California Division of Agricultural Sciences, 1974): 197.

108 *Although he was not even born at the time of the Banner's creation:* Interview with Miles Reiter (3 December 1998, Naomi Hirahara and Teru Kanazawa Sheehan, Watsonville, Calif.).

109 *Yellows was a new disease:* Harold A. Johnson, Jr., "The Contributions of Private Strawberry Breeders," *HortScience* 25, no. 8 (August 1990): 901.

109 *According to strawberry expert Stephen Wilhelm:* Stephen Wilhelm and James E. Sagen, *A History of the Strawberry: From Ancient Gardens to Modern Markets* (University of California Division of Agricultural Sciences, 1974): 214.

109 *To combat such diseases:* George M. Darrow, *The Strawberry: History, Breeding and Physiology* (New York: Holt, Rinehart and Winston, 1966): 227.

109 *Those who worked with Dr. Thomas:* Interview with Harold Johnson (7 November 1999, Naomi Hirahara, Watsonville, Calif.).

109 *But he had an even better education right on the farm:* Stephen Wilhelm and James E. Sagen, *A History of the Strawberry: From Ancient Gardens to Modern Markets* (University of California Division of Agricultural Sciences, 1974): 217.

109 *According to a U.S. Department of Agriculture report:* Diane Bertelson, "The U.S. Strawberry Industry, Commercial Agricultural Division," Economic Research Service, U.S. Department of Agriculture, Statistical Bulletin, no. 914, accessed 27 January 1999.

109 *Harold Johnson, a former breeder for Driscoll:* Interview with Harold Johnson (5 November 1999, Naomi Hirahara and Teru Kanazawa Sheehan, Watsonville, Calif.).

110 *One of those who also shared Ned's passion:* Harold A. Johnson Jr., "The Contributions of Private Strawberry Breeders," *HortScience* 25, no. 8 (August 1990): 901.

110 *In 1944 Ned Driscoll had launched:* Harold A. Johnson Jr., "The Contributions of Private Strawberry Breeders," *HortScience* 25, no. 8 (August 1990): 901; Stephen Wilhelm and James E. Sagen, *A History of the Strawberry: From Ancient Gardens to Modern Markets* (University of California Division of Agricultural Sciences, 1974): 230.

Chapter Five: Ned Driscoll and the Z5-A

112 *Harold Johnson remembers:* Interview with Harold Johnson (5 November 1999).

112 *As described in Darrow's book:* George M. Darrow, *The Strawberry: History, Breeding and Physiology* (New York: Holt, Rinehart and Winston, 1966): 230.

112 *It was the Z5-A that convinced the Uyematsu family:* Interview with Richard Uyematsu.

Chapter Six: Watsonville

113 *It was originally called Bodfish Canyon Road:* Betty Lewis, *Watsonville: Memories That Linger,* vol. 2 (Santa Cruz, Calif.: Valley Publishers, 1980): 24.

113 *Charles Iwami's father, Yasutaro:* Interview with Charles Iwami (10 November 1999, Naomi Hirahara, Watsonville, Calif.).

Further Reading

Sandy Lydon, *The Japanese in the Monterey Bay Region: A Brief History* (Capitola, Calif.: Capitola Book Company, 1997).

Chapter Seven: Braceros

116 *The* bracero *program provided a steady labor force:* "Year by Year: Southern California Chronology," *Los Angeles Times Magazine,* 10 January 1999; Miriam J. Wells, *Strawberry Fields: Politics, Class and Work in California Agriculture* (Ithaca, New York: Cornell University Press, 1996): 51–60.

117 *As a result, class action lawsuits:* Ken Ellingwood, "Braceros Demand Lost Legacy," *Los Angeles Times,* 15 November 1999; James F. Smith, "Ex-Migrants Sought for Class-Action," *Los Angeles Times,* 15 March 2001.

FURTHER READING

Henry Pope Anderson, *The* Bracero *Program in California* (New York: Arno Press, 1976).

Kitty Calavita, *Inside the State: The Bracero Program, Immigration, and the I.N.S.* (New York: Routledge, 1992).

Deborah Cohen, "Caught in the Middle: The Mexican State's Relationship with the United States and Its Own Citizen-Workers, 1942–1954," *Journal of American Ethnic History* 20, no. 3 (Spring 2001).

Richard B. Craig, *The Bracero Program* (Austin, Texas: University of Texas Press, 1971).

CHAPTER EIGHT: PROTECTING THE ROOTS

118 *Methyl bromide is a manmade chemical:* Jeff Wheelwright, "The Berry and the Poison," *Smithsonian* 27, no. 9: 40.

118 *You had to be especially careful with land:* Stephen Wilhelm and James E. Sagen, *A History of the Strawberry: From Ancient Gardens to Modern Markets* (University of California Division of Agricultural Sciences, 1974): 218.

119 *After it was shown to deplete the ozone layer:* Jeff Wheelwright, "The Berry and the Poison," *Smithsonian* 27, no. 9: #40.

FURTHER READING

Stephen Wilhelm, "Soil Fumigation," *The Grower*, March 1952.

CHAPTER NINE: BROKEN BOTTLES

121 *After he died, one of his obituaries:* "Kiyoshi Hirasaki, 63, Prominent Gilroy Grower and Community Leader Dies," *Nichi Bei Times*, 25 December 1963:1.

CHAPTER ELEVEN: MOTHERS, DAUGHTERS, FATHERS, AND PREGNANT PLANTS

124 *We would take the runners*: George M. Darrow, *The Strawberry: History, Breeding and Physiology* (New York: Holt, Rinehart and Winston, 1966): 329; Interview with Harold Johnson (7 December 1999).

124 *Al Amorao, born in the Philippines:* Interview with Amado "Al" Amorao (30 July 2002, Naomi Hirahara, Oxnard, Calif.).

125 *Harold Johnson had been hard at work in developing the* G3: Interview with Harold Johnson (5 December 1999).

CHAPTER TWELVE: CULTURAL PRACTICES

127 *Sho Kobara describes it:* Interview with Shoichi Kobara.

128 *Everyone had his or her particular way:* Miriam J. Wells, *Strawberry Fields: Politics, Class and Work in California Agriculture* (Ithaca, New York: Cornell University Press, 1996): 131.

128 *We'd place plastic tarps:* Ibid., 176.

CHAPTER THIRTEEN: HEADING SOUTH

130 *Decades later:* Interview with John German (5 November 1999, Naomi Hirahara and Teru Kanazawa Sheehan, Watsonville, Calif.).

PART FOUR: SUMMER

Chapter Two: Oxnard

143 *According to community historian Yoshio Fukuyama:* Yoshio Fukuyama, "The Japanese in Oxnard, California, 1898–1945" and "Citizens Apart: A History of the Japanese in Ventura County," *Ventura County Historical Society Quarterly* 39, no. 4; 40 no. 1, 1994).

143 *By the early 1900s, Chinese immigrants:* Margaret Jennings, "The Chinese in Ventura County," *Ventury County Historical Society Quarterly* 29, no. 3 (Spring 1984): 6–7.

143 *A Chinese man named Ung Hing:* "Ortega Adobe Historic Residence" (brochure), Ventura, Calif.: City of San Buenaventura Community Services Department, n.d.

144 *In addition to a large percentage of their wages going to the labor contractor:* Juan Gomez-Quinones, "The First Steps: Chicano Labor Conflict and Organizing 1900–1920," *Aztlan Chicano Journal of the Social Sciences and Arts* 3 (Spring 1972): 24.

144 *One of the leaders of the labor association:* Yoshio Fukuyama, "The Japanese in Oxnard, California, 1898-1945" and "Citizens Apart: A History of the Japanese in Ventura County," *Ventura County Historical Society Quarterly* 39, no. 4; 40, no. 1, 1994).

144 *He was also opposed to gambling in town:* "Origin of Oxnard Nisei Church Recalled by Present Minister," *Ventura County (California) Star-Free Press*, 14 October 1948.

145 *Yonejiro explains why:* Interview with Yonejiro Ito (5 February 1999, Naomi Hirahara and Teru Kanazawa Sheehan, Oxnard, Calif.).

145 *He initially specialized:* Interview with Paul Murata (23 June 1989, Vernon Takeshita, Calif.)

FURTHER READING

Tomas Almaguer, "Racial Domination and Class Conflict in Capitalist Agriculture," in *Working People of California*, Daniel Cornford, ed. (Berkeley: University of California Press, 1995): 183–207.

Verna Bloom, "Oxnard: A Social History of the Early Years," *The Ventura County Historical Society Quarterly* 4, no. 2 (February 1959): 13–20.

Ventura County Historical Society Quarterly 23, no. 4 (Summer 1978).

CHAPTER THREE: E.F. DRISCOLL FARMING TRUST

147 *The Milias, the oldest and largest hotel in Gilroy:* "Gilroy Historic District Walking Tour–Monterey Street" (pamphlet), Gilroy Downtown Development Corp., Gilroy Visitor's Bureau, Gilroy Historical Society, City of Gilroy Historical Museum, n.d.

CHAPTER FOUR: DRISCOLL STRAWBERRY ASSOCIATES

149 *From the 1990s:* Interview with John German.

149 *Modern berry marketing goes back to 1912:* Stephen Wilhelm and James E. Sagen, *A History of the Strawberry: From Ancient Gardens to Modern Markets* (University of California Division of Agricultural Sciences, 1974): 198.

149 *At its core:* Miriam J. Wells, *Strawberry Fields: Politics, Class and Work in California Agriculture* (Ithaca, New York: Cornell University Press, 1996): 126.

149 *In February 1953, Driscoll Strawberry Associates:* Articles of Incorporation of Driscoll Strawberry Associates, Inc., Filed: 20 February 1953.

CHAPTER FIVE: REMOVING LEMONS

152 *Lemons were introduced:* Dirk Werkman, "Agricultural History Harvest Rich in County," *Ventura County Star*, 29 April 1978.

152 *In 1927:* Ventura County Annual Report (Crop Statistics), Agricultural Commissioner, 1947; *Ventura County Annual Report* (Crop Statistics), Agricultural Commissioner, 1949.

Chapter Six: Sharecropping vs. Subcontracting

156 *Some farmers were hit with lawsuits:* Miriam J. Wells, *Strawberry Fields: Politics, Class and Work in California Agriculture* (Ithaca, New York: Cornell University Press, 1996): 245.

Chapter Seven: Pitching Nets

157 *In terms of progressive culture:* Interview with Harold Johnson (5 December 2003).

158 *Pomology involves:* Interview with Victor Voth (23 June 1989, Vernon Takeshita).

Further Reading

Laura Avery, "Better Strawberries Through Science," *Santa Monica Mirror*, vol. 2:27: 20–26.

Chapter Eight: New Community

160 *It was built on East Sixth Street:* "Oxnard Buddhist Church: 50 Years History, 1929–1979" (booklet), (Oxnard, Calif.: Oxnard Buddhist Church, 17 November 1979).

160 *Located along a road lined with eucalyptus trees:* "History of Oxnard Buddhist Temple: Update 1980–1999" (booklet), (Oxnard, Calif.: Oxnard Buddhist Temple and Buddhist Women's Association, September 1999).

160 *Hanako:* Tsujio Kato, "President's Message–January 1995," *Four Seasons: Ventura County Japanese Citizens League Quarterly Newsletter* 14, no. 1.

161 *On October 27, 1936:* "Dispute Over Control of Seventh Street Center Calm," *Oxnard Press-Courier*, 29 October 1949.

161 *In March 1942:* "Jap-American Plea Backed," *Oxnard Press-Courier*, 31 October 1949.

161 *In March 1951:* "Japanese-Americans Win Right to 7th Street Recreation Center," *Oxnard Press-Courier*, 21 March 1951.

Chapter Nine: Right Variety

162 *Johnson had a summer-planted variety called the D-4:* Interview with Harold Johnson (5 December 1999).

163 *To create a variety, you first cross the seeds:* Interview with Amado "Al" Amorao.

Chapter Ten: Union and Protests

164 *What no strawberry grower wants to talk about:* Miriam J. Wells, *Strawberry Fields: Politics, Class and Work in California Agriculture* (Ithaca, New York: Cornell University Press, 1996): 74–107.

Chapter Twelve: Entering the Circle

171 *Ned Driscoll, called "Mr. Strawberry of California":* "Peers Warmly Remember 'Mr. Strawberry' Driscoll," *The Packer*, 14 November 1981: 3A.

Chapter Thirteen: Manabi Farms and Long-Stem Strawberries

173 *Born in Monterey Park in 1939:* Interview with Cecil Martinez (28 October 1999, Naomi Hirahara and Teru Kanazawa Sheehan, Oxnard, Calif.).

Chapter Fourteen: Longest Partnership, New Passions

176 *My daughter, Marcia, says:* Interview with Marcia (Hirasaki) Messinger (5 December 1998, San Francisco).

177 *Since she was raised on a carnation farm:* Interview with Sumi (Iwata) Hirasaki.

Further Reading

"Hirasakis' Grant to Yu-Ai Kai Purchases Vans for Seniors," *Hokubei Mainichi*, 3 May 1984: 2.

"Oxnard Grower Gives $25,000 to Museum," *The Rafu Shimpo*, 8 December 1986.

Chapter Fifteen: Father and Son

180 *My daughter, Marcia:* Interview with Marcia (Hirasaki) Messinger.

Chapter Sixteen: Strawberry Diplomacy

181 *The late Tsujio Kato started the California Strawberry Festival:* "Planting Seeds–How the California Strawberry Festival Began," California Strawberry Festival Web site, www.strawberry-fest.org, accessed 23 March 2000.

181 *In 1989, shortly after I stepped down:* "America's Strawberry: Fruit of Our Labor" (dinner booklet), Japanese American National Museum, 1989.

Further Reading

Yoshimi Ishikawa (Eve Zimmerman, trans.), *Strawberry Road* (Tokyo: Kodansha International, 1991).

David Mas Masumoto, *Epitaph for a Peach: Four Seasons on My Family Farm* (San Francisco: HarperSanFrancisco, 1995).

Valerie Matsumoto, *Farming the Home Place: A Japanese American Community in California, 1919–1982* (Ithaca, New York: Cornell University Press, 1993).